Civil Engineering Standard Method of Measurement

Examples

CESMM4

Civil Engineering Standard Method of Measurement Examples

Institution of Civil Engineers

Published by ICE Publishing, One Great George Street, Westminster, London SW1P 3AA

Full details of ICE Publishing sales representatives and distributors can be found at:
www.icevirtuallibrary.com/printbooksales

Also available from ICE Publishing:

CESMM4 Civil Engineering Standard Method of Measurement
Institution of Civil Engineers. ISBN 978-0-7277-5751-7
CESMM4 Carbon & Price Book 2013
Franklin & Andrews and Institution of Civil Engineers. ISBN 978-0-7277-5812-5

www.icevirtuallibrary.com

A catalogue record for this book is available from the British Library

ISBN 978-0-7277-5759-3

ICE Publishing is a division of Thomas Telford Ltd, a wholly owned subsidiary of the Institution of Civil Engineers (ICE).

Senior Commissioning Editor: Gavin Jamieson
Production Editor: Rebecca Taylor
Market Development Executive: Elizabeth Hobson

Typeset by Academic + Technical, Bristol
Printed and bound in Great Britain by TJ International Ltd, Padstow

Contents

Introduction

The Work Classification divides the work which is covered by the Civil Engineering Standard Method of Measurement (CESMM) into 26 classes lettered A to Z. Each class contains three types of information: an 'includes and excludes' list, a classification table and a table of rules.

The includes and excludes list is given at the head of each class. It tells the user of CESMM4 which general types of work are included in a class and which classes cover other similar work which is excluded from that class. In some cases, for example in classes A, C, D, I and V, the scope or coverage of the items in a class is also given. These lists should not be regarded merely as an index to CESMM4; they are important to the interpretation of the coverage of the bill items generated by the classes. Clearly the lists do not set out to be comprehensive; they do not mention everything which is included or everything which is excluded.

The classification table is the tabulation of the work components covered by a class, divided into the three divisions as described in section 3. The horizontal lines in the tabulation indicate which lists of features from one division apply to which features in the other divisions. This is usually straightforward, but must be given close attention where the lines are at different levels in the different divisions. Headings are given in some lists of features and are printed in *italics* in CESMM4. They should be included in item descriptions in all cases where they would not duplicate information. For example, the heading 'Cement to BS 12 or BS 146', is obviously essential in the descriptions for items F 1-3 * 1-4 covering provision of concrete as otherwise the information about cement is not given.

In many places the classification table uses the word, 'stated', in phrases; such as, 'Formwork: stated surface features', and, 'Width: stated exceeding 300 mm'. Written in full these phrases might become: 'This item classification is for formwork which has a particular surface feature. Descriptions of items in this classification shall state the particular surface feature required' and 'This item classification is for things the width of which exceeds 300 mm. Descriptions of items in this classification shall state the actual width of the things required'.

The rules on the right-hand pages are as important as the classification tables. In one sense they are more important as sometimes they overrule the classification table. The rules are arranged alongside the sections of the classification to which they apply. This is indicated by the horizontal lines which align from the left- to the right-hand pages. Rules printed above a double line apply to all items in a class (see paragraph 3.11).

The terms which are printed in *italics* in the rules are those which are CESMM4 EXAMPLES taken directly from the classification table. This style of printing is adopted as an aid to cross-reference between the tables and the rules; it has no effect on interpretation of the rules.

CESMM4 uses some untraditional terms. They are adopted to comply with British Standards or to keep up with the move to standardise units and terms under the general umbrella of metrication. Thus pipes have a bore not a diameter, because the bore is the diameter of the hole down the middle, and cannot be confused with the outside diameter of the pipe. Mass is the measure of the quantity of matter; weight is no longer used. CESMM4 refers to weight in only one place where it was considered that the alternative phrase 'piece mass' would be totally unfamiliar. The abbreviation for number is 'nr'.

Example bill pages are given in this book for each class in the Work Classification. The examples are not taken from actual contracts. They can be used as a guide to the layout and style of bills and bill items compiled using CESMM4. The example bill items, not being related to a particular job, show less non-standard description amplifying the basic descriptions than is given in real bills. Similarly, in order not to imply that particular specification details are recommended, item descriptions in the example bills frequently refer to hypothetical specification clauses by a clause number or to details on hypothetical drawings. This procedure is permitted in real bills by paragraph 5.12 but it is not adopted in them to the same extent as it has been in the example bill pages.

The examples use the code numbers in the Work Classification as item numbers. This practice is recommended but is not a requirement of CESMM4. It is adopted at the discretion of the bill compiler in accordance with paragraph 4.3.

The examples do not cover all the items which could be generated by CESMM4 or even all the items which might occur in one bill. They give hypothetical items which demonstrate those applications of the rules in CESMM4 which are novel or would benefit from demonstration for other reasons. The items are laid out as if they were pages from a bill to demonstrate layout, numbering and the use of headings. The examples illustrate the alternative procedures open to bill compilers where CESMM4 permits alternatives. Compilers of real bills should try to be consistent, not to demonstrate all the possible alternatives as the examples do.

Each heading and item description in the example bill pages ends with a full stop. This is a helpful discipline when there are two or more headings at the top of one bill page. The full stops help to relate the headings to the lines drawn across the description column which show which items apply to each heading (see paragraph 5.9). Within item descriptions, a semicolon has been used to separate basic from additional description.

Diagrams

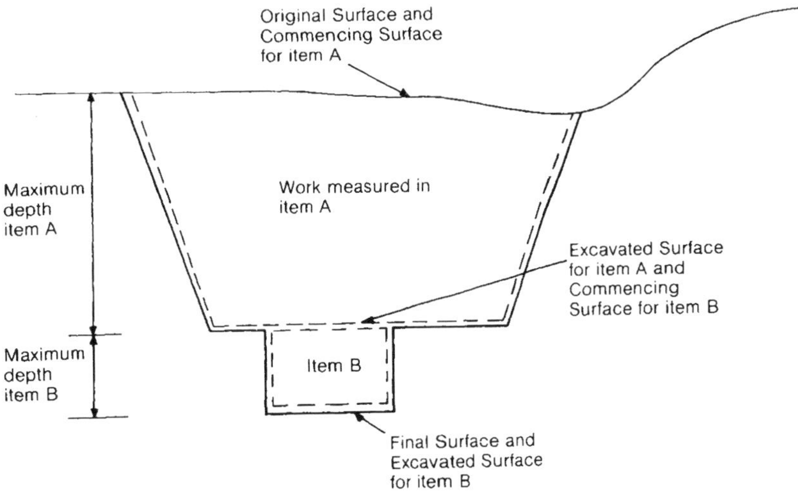

Figure 1. Application of the definitions of the four surfaces given in paragraphs 1.10–1.13. The Excavated Surface for one item becomes the Commencing Surface for the next item if excavation is measured in more than one stage (see also paragraph 5.21)

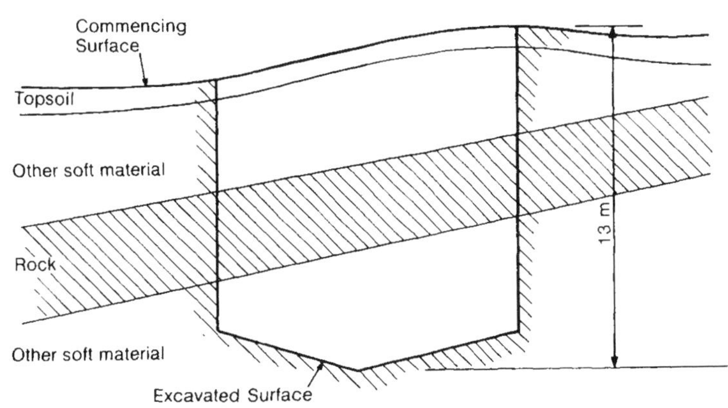

Figure 2. Three items are required for this excavation. All can be described as 'maximum depth 10–15 m'. Definition rules 1.12 and 1.13 do not require intermediate surfaces to be identified

CLASS I: PIPEWORK – PIPES

Includes: **Provision, laying and jointing of pipes**
Excavating and backfilling pipe trenches
Excludes: **Work included in classes J, K, L and Y**
Piped building services (included in class Z)

FIRST DIVISION		SECOND DIVISION	THIRD DIVISION
1 Clay pipes	m	1 Nominal bore: not exceeding 200 mm	1 Not in trenches
2 Concrete pipes	m	2 200–300 mm	2 In trenches, depth: not exceeding 1·5 m
3 Iron pipes	m	3 300–600 mm	3 1·5–2 m
4 Steel pipes	m	4 600–900 mm	4 2–2·5 m
5 Polyvinyl chloride pipes	m	5 900–1200 mm	5 2·5–3 m
6 Glass reinforced plastic pipes	m	6 1200–1500 mm	6 3–3·5 m
7 High density polyethylene pipes	m	7 1500–1800 mm	7 3·5–4 m
8 Medium density polyethylene pipes	m	8 exceeding 1800 mm	8 exceeding 4 m

Figure 3. Classification table for pipes in class I. This is the simplest table in CESMM3 and shows clearly how the three divisions of classification combine to produce brief descriptions and code numbers for groups of components of civil engineering works. In this case the brief descriptions are amplified in bills of quantities by more specific information given in accordance with the additional description rules in class I

CLASS C

MEASUREMENT RULES	DEFINITION RULES	COVERAGE RULES	ADDITIONAL DESCRIPTION RULES
M1 The Commencing Surface adopted in the preparation of the Bill of Quantities shall be adopted for the measurement of the completed work. **M2** The depths of *grout holes*, holes for *ground anchorages* and *drains* shall be measured along the holes irrespective of inclination.	**D1** Drilling and excavation for work in this class shall be deemed to be in *material other than rock or artificial hard material* unless otherwise stated in item descriptions.	**C1** Items for work in this class shall be deemed to include disposal of excavated material and removal of dead services.	
M3 Drilling through previously grouted holes in the course of stage grouting shall not be measured. Where holes are expressly required to be extended, the number of holes shall be *measured and drilling* through previously grouted holes shall be measured as *drilling through rock or artificial hard material.* **M4** The *number of stages* measured shall be the total number of grouting stages expressly required.			**A1** The diameters of holes shall be stated in item descriptions for *drilling and driving for grout holes* and *grout holes.*

Figure 4. The layout of the classified rules in CESMM3: note the different style of each of the four types of rule, the horizontal alignment and the use of the double horizontal line to separate rules of general application to the class

Section A. List of principal quantities
Section B. Preamble
Section C. Daywork Schedule
Section D. Work items
- Part 1. General items
- Part 2. Outfall
- Part 3. Plowden treatment works
- Part 4. West branch sewers
- Part 5. East branch sewers
- Part 6. Whettleton pumping station

Section E. Grand Summary

Figure 5. Example of the standardised sequence of contents of a bill of quantities which results from application of paragraphs 5.2 and 5.8. Sections of the Bill are identified by the letters A to E to distinguish them from the numbered parts into which the work items themselves are divided

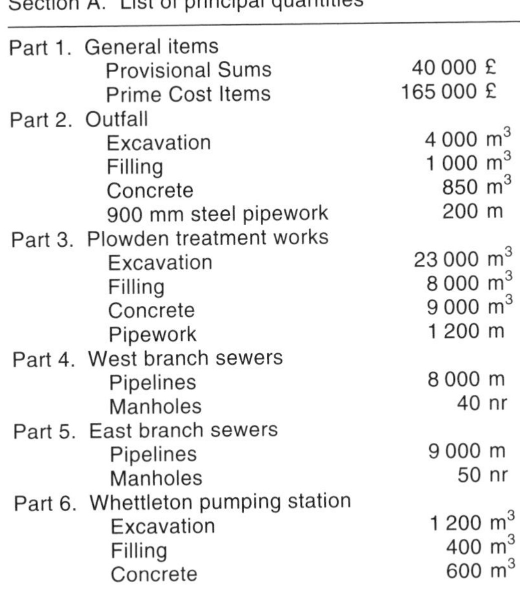

Section A. List of principal quantities	
Part 1. General items	
Provisional Sums	40 000 £
Prime Cost Items	165 000 £
Part 2. Outfall	
Excavation	4 000 m^3
Filling	1 000 m^3
Concrete	850 m^3
900 mm steel pipework	200 m
Part 3. Plowden treatment works	
Excavation	23 000 m^3
Filling	8 000 m^3
Concrete	9 000 m^3
Pipework	1 200 m
Part 4. West branch sewers	
Pipelines	8 000 m
Manholes	40 nr
Part 5. East branch sewers	
Pipelines	9 000 m
Manholes	50 nr
Part 6. Whettleton pumping station	
Excavation	1 200 m^3
Filling	400 m^3
Concrete	600 m^3

Figure 6. Example of a list of principal quantities compiled to comply with paragraph 5.3

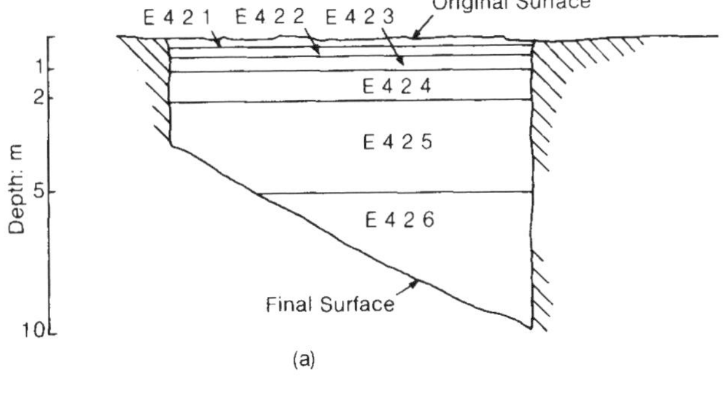

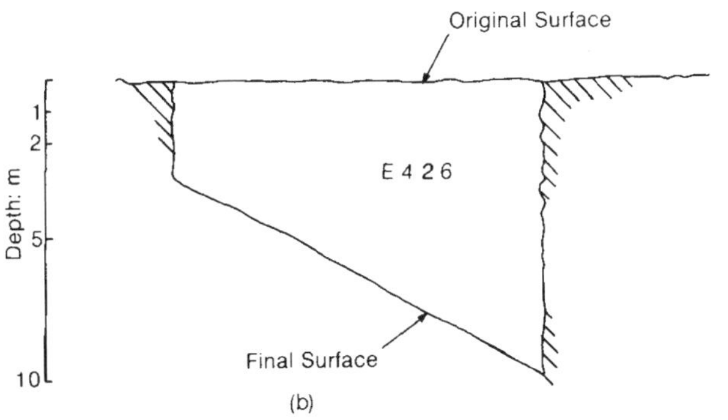

Figure 7. CESMM3 does not divide excavation into depth bands (a), but according to total depth (b)

Figure 8. Column layout described in paragraph 5.22 and typing capacity of the columns when a typewriter producing ten characters to the inch is used

Number	Item description	Unit	Quantity	Rate	Amount £	p
X999.99	Description line max 34 characters	sum	9999999	9999999	9999999	99
		m2	99999.9	0.99		

Figure 9. Time based graph showing how the payment of the Adjustment Item sum relates to the tendered value of work items. It illustrates the effect of paragraph 6.4

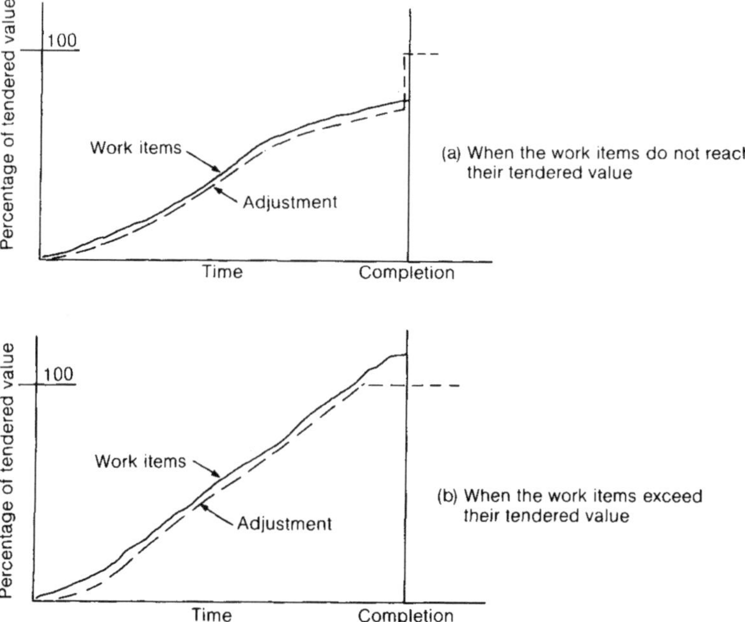

Figure 10. All prices against items in bills of quantities compiled using CESMM3 have an assumed relationship to cost. This relationship is either to quantities which can be observed in the physical work itself (quantity-proportional unit rates), to time (Time-Related Charges) or to neither quantity nor time (Fixed Charges). This figure shows where the items which embody these three relationships are to be found in CESMM3

Work items
General items
Related to quantites
Relationship specified
Specified requirements
Method-Related Charges
Fixed
Time related
Quantity related
Fixed
Time related

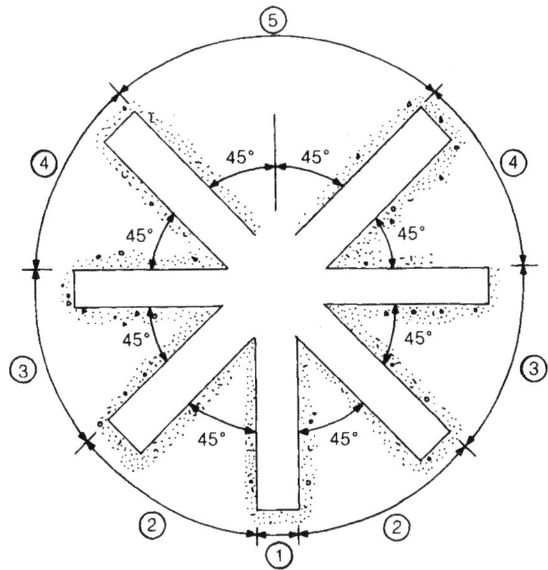

Figure 11. Zones of inclination for grout hole drilling and driving given in the second division of class C, item codes C 1–3 1–5*. Notice the precise boundaries for the zones. For example, drilling at 45° to the vertical upwards is in zone 4, drilling at 45° to the vertical upwards is in zone 5

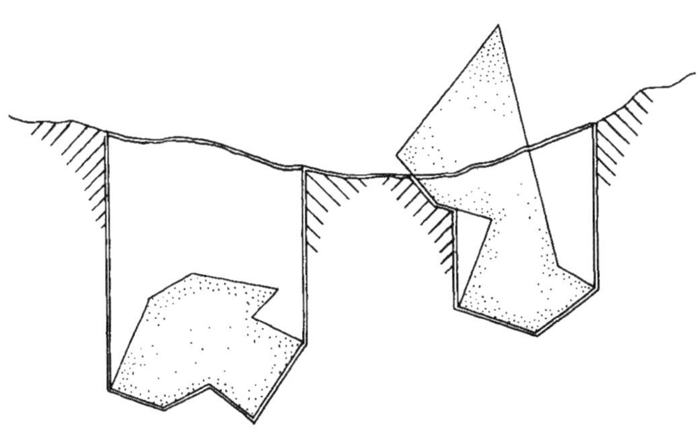

Figure 12. These ridiculously shaped structures, which are to be constructed after excavation for them has been carried out, demonstrate the application of rules M6 and M16. The volume measured for excavation is the volume bounded by the double line. The volume measured for backfilling, if required, would be the part of this volume not occupied by the finished structure

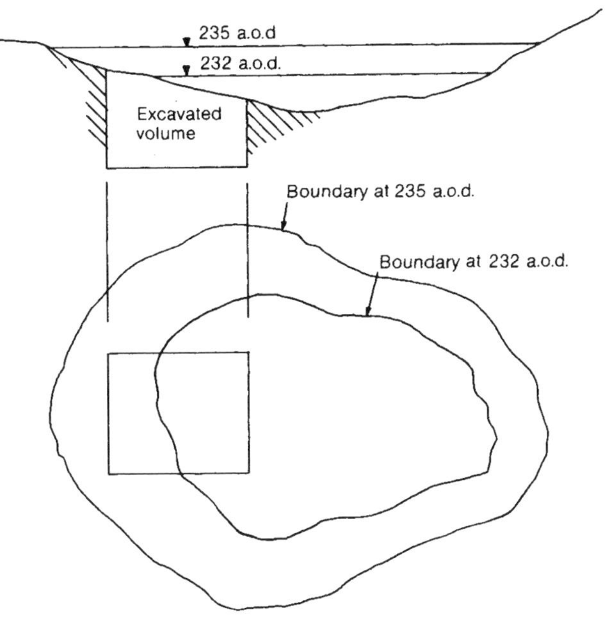

Figure 13. The volume measured for excavation below water is that which is below water at the higher anticipated level

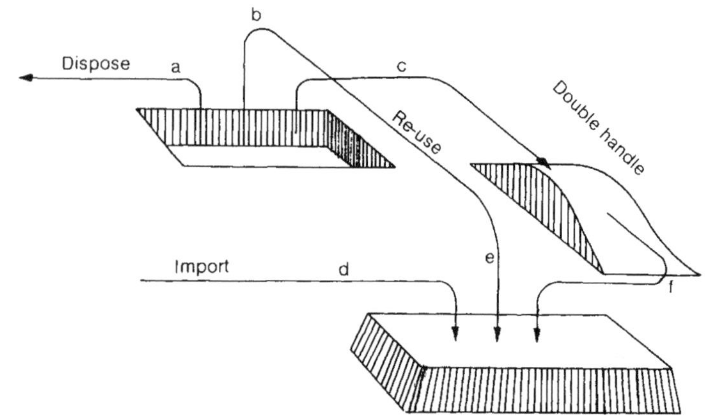

Figure 14. The general case of the cut and fill operation. When a composite earthmoving operation such as that represented here is to be measured, care must be taken to ensure that enough quantities are measured in the field to enable the quantities against the bill items to be calculated

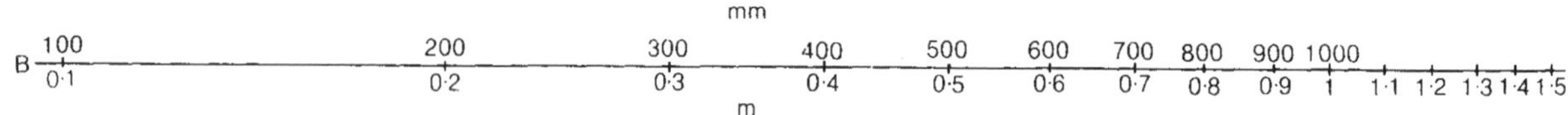

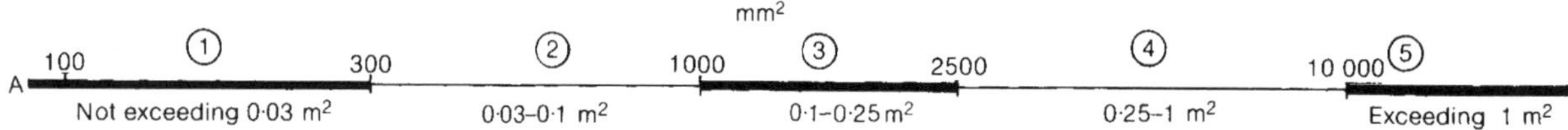

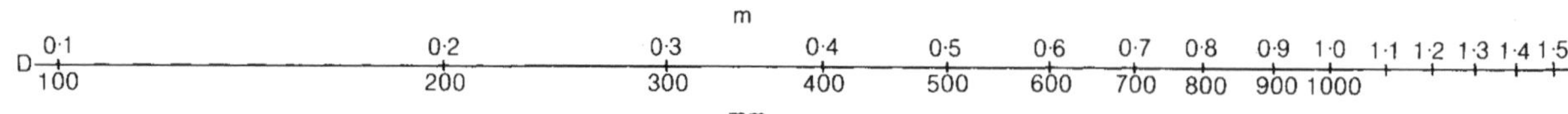

Figure 15. Nomogram for areas of cross-section of rectangular columns, beams, piers and casing to metal sections. If a straight edge is placed from the breadth dimension on scale B to the depth on scale D, the intersection on scale A shows the range from CESMM3 in which the cross-sectional area of the component occurs. If the intersection is very close to a range boundary, it is necessary to check by calculation

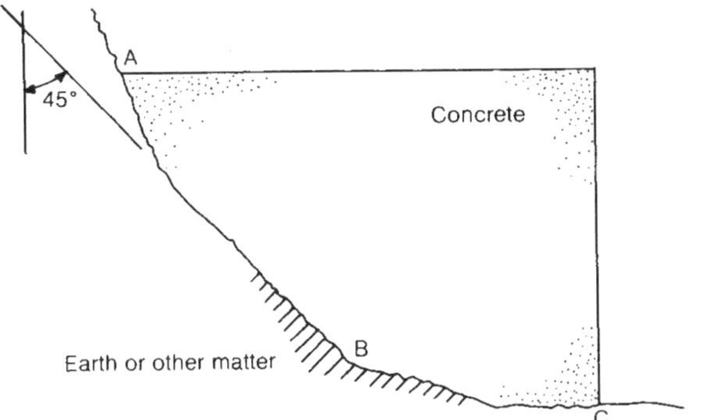

Figure 16. Illustration of rule M2(e). Formwork is measured for the surface AB but not for the surface BC

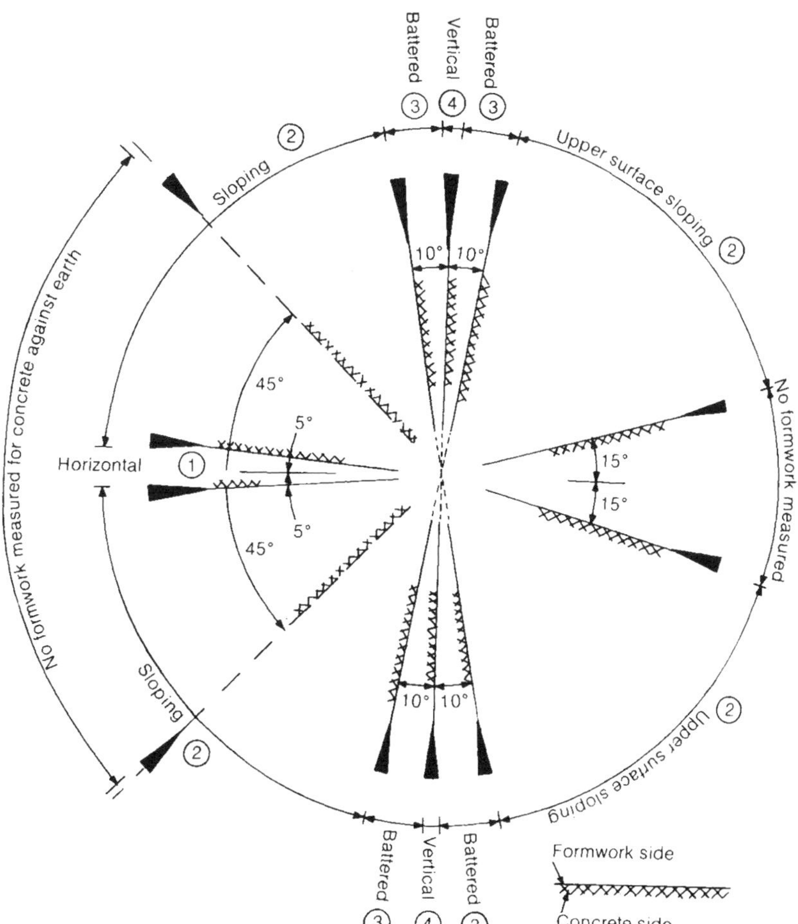

Figure 17. Inclination zones for plane formwork defined in rules M2(e), M3, D1 and A2. Note the precise boundaries of the zones. For example, an inclination of 10° to the vertical is in zone 3; an inclination of 10½° is in zone 2

Figure 18. Inserts classified according to rule A15. The four types of insert are shown illustrating their different effects on wall formwork. The figure viewed on its side illustrates the same points for slab formwork

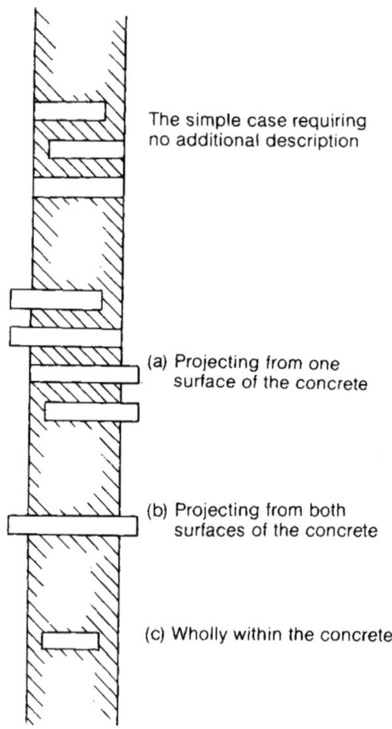

Figure 19. Measurement of pipe trench depths. The depths of pipe trenches are given in zones (measured to pipe invert level). Thus any variation in trench depth at remeasurement shows itself as a change to the lengths of pipe which occur in each depth zone

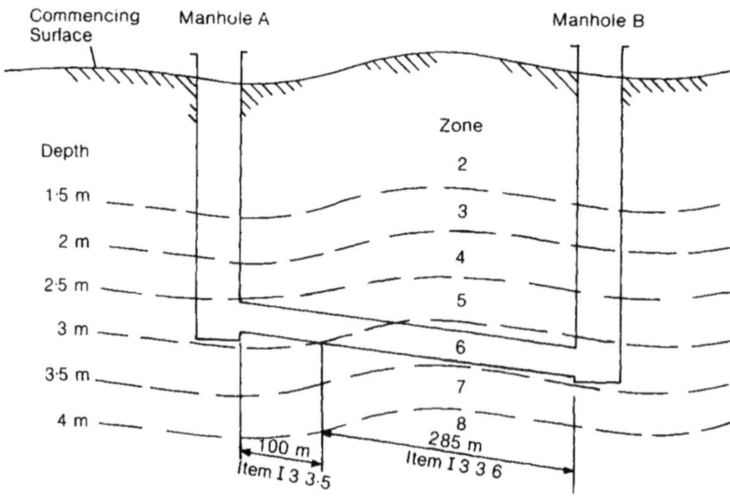

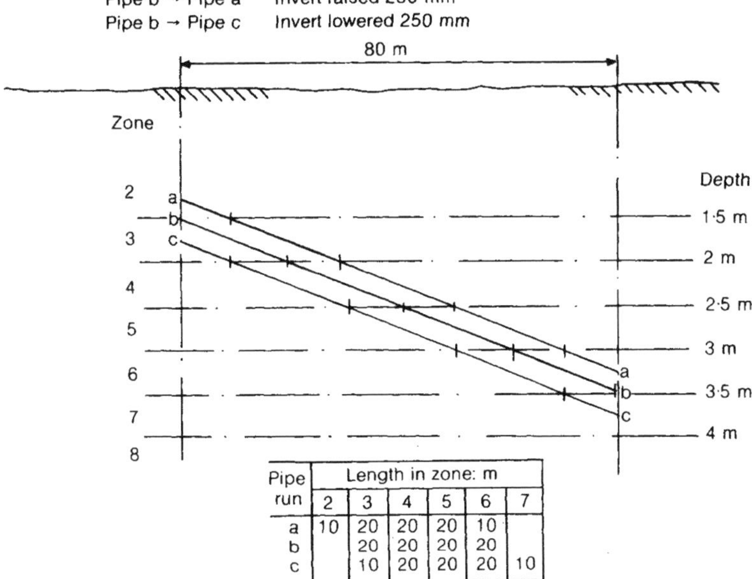

Pipe run	Length in zone: m					
	2	3	4	5	6	7
a	10	20	20	20	10	
b		20	20	20	20	
c		10	20	20	20	10

Figure 20. Simplified example showing how remeasurement of a pipe run in depth zones deals with variations in pipe trench depths

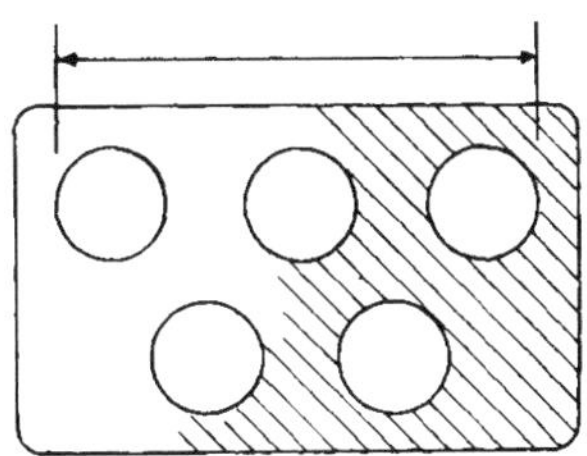

Figure 21. Dimension used in calculating nominal trench widths for multiple bore ducts and twinned pipes. This is described in rules D1 and D7 of class K and A3, M2 and D1 of class L. The dimension is referred to as the 'distance between the inside faces of the outer pipe walls'

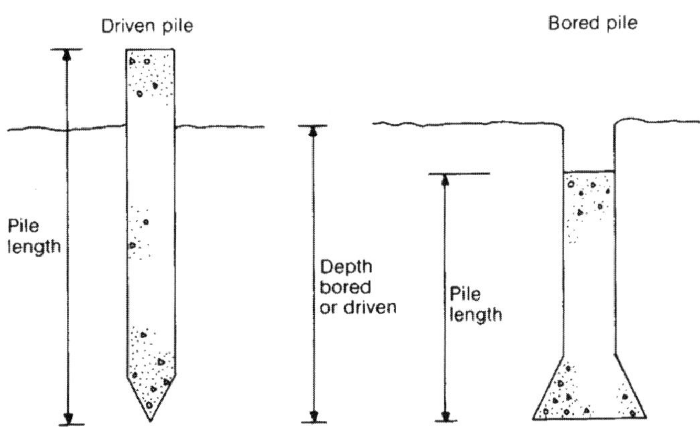

Figure 22. Measurement of piles. At least two bill items are given for each group of preformed concrete, timber or isolated steel piles: number of piles of a stated length and depth driven. At least three bill items are given for each group of cast in place piles: number of piles, concreted length and depth bored or driven stating the depth of the deepest pile in the group. These items are generated by the third division of classification of class P and associated rules

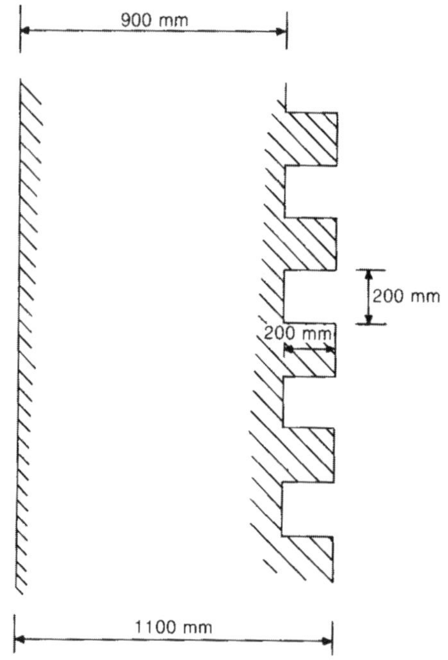

Figure 23. A wall with these surface features could be considered as of thickness 900 mm with 200 mm × 200 mm projections or as of thickness 1100 mm with 200 mm × 200 mm rebates. Additional description must be given when work such as this – straddling the boundary of two parts of the Work Classification – is to be measured. This is the effect of paragraph 5.13. The wall in the figure should be identified using additional description to avoid uncertainty

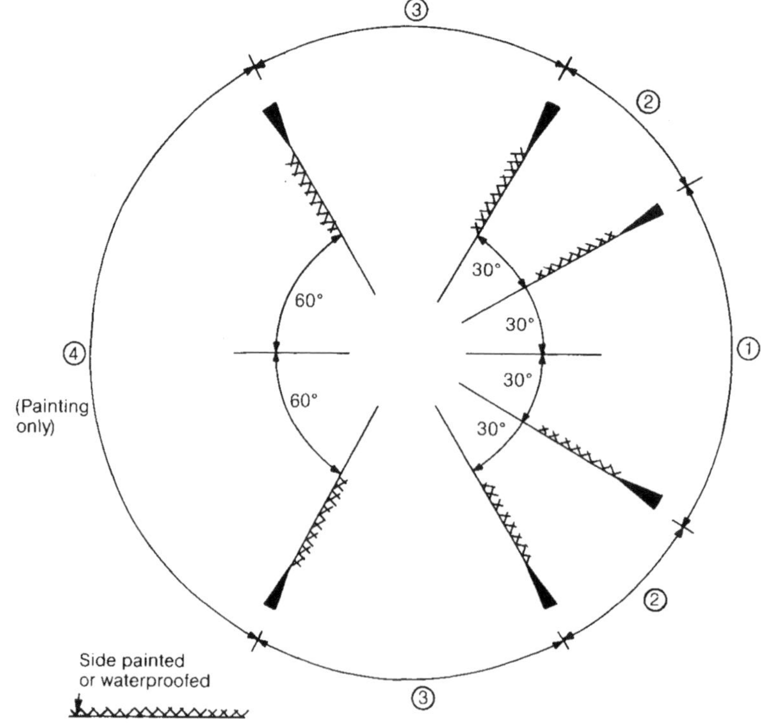

Figure 24. Inclination zones used for classification of painting and waterproofing of plane surfaces of width exceeding 1 m as given in the third division of classes V and W. (See also rule M2.) Note the precise boundaries of the zones. For example, a soffit surface inclined at 60° to the horizontal is classed in zone 4; one inclined at 61° is classed in zone 3

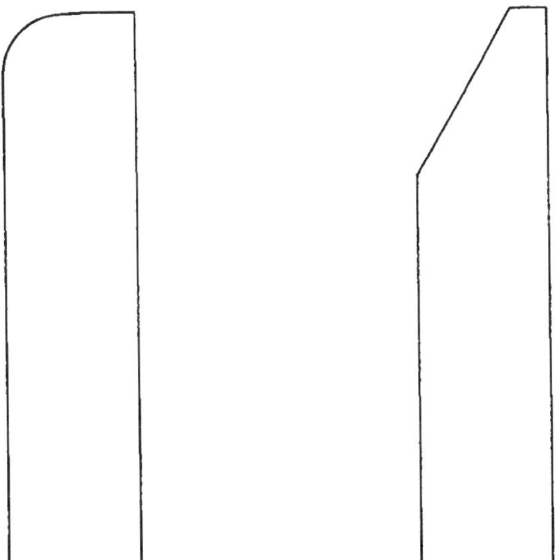

Figure 25. Skirting boards are measured as a single item where they have the same characteristics but different shapes (rule A4). These skirting boards have the same dimensions but two different shapes

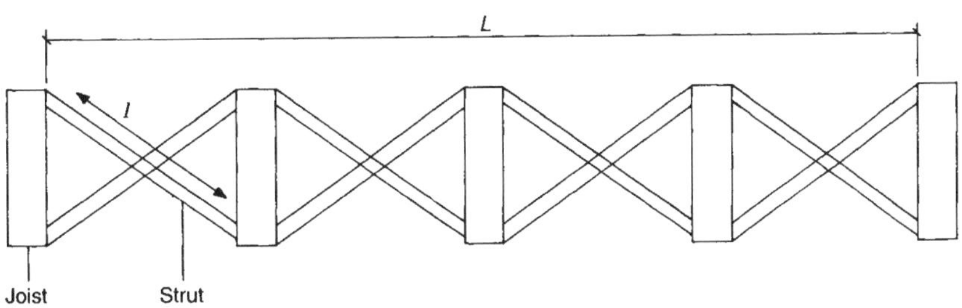

Figure 26. Rule M3 simplifies the measurement of diagonal strutting. The length measured is 2 × L rather than 8 × *l*

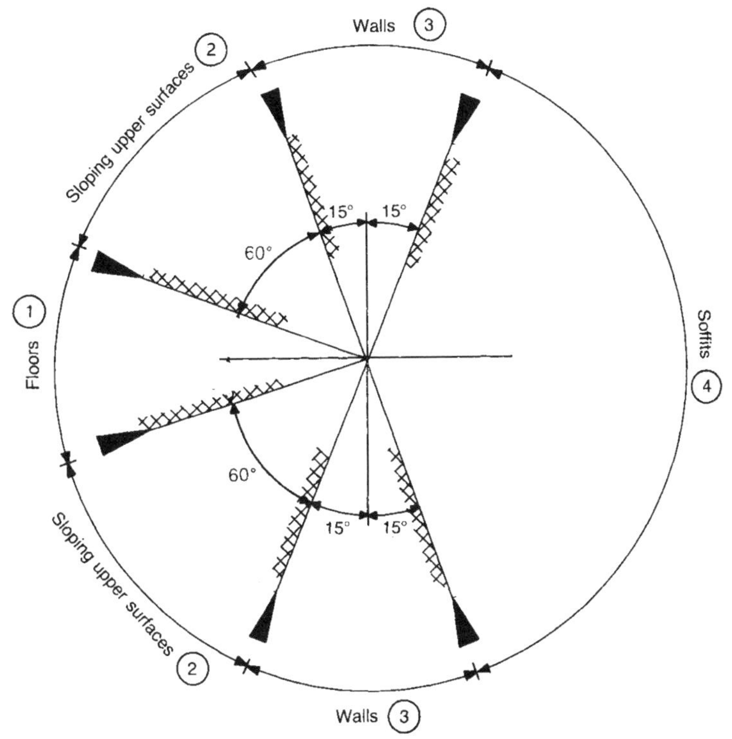

Figure 27. Inclination zones used for classification of boarding and insulation (rule D4) and surface finishes, linings and partitions exceeding 1 m wide (rule D7). Note the precise boundaries of the zones. For example, a sloping upper surface inclined at 75° to the vertical is classed in zone 2; one inclined at 76° is classed in zone 1

Example bill of quantities pages

Number	Item description	Unit	Quantity	Rate	Amount	
					£	p
	PART 1: GENERAL ITEMS.					
	Contractual requirements.					
A110	Performance bond.	sum				
A120	Insurance of the Works.	sum				
A130	Third party insurance.	sum				
	Specified requirements.					
A211.1	Establishment and removal of offices for the Engineer's staff.	sum				
A211.2	Maintenance of offices for the Engineer's staff.	sum				
A211.3	Maintenance of offices for the Engineer's staff after issue of the Substantial Completion Certificate.	wk	20			
A212.1	Establishment and removal of laboratories for the Engineer's staff.	sum				
A212.2	Maintenance of laboratories for the Engineer's staff.	sum				
A221.1	Transport vehicle; as Specification Clause 184.9.	wk	208			
A221.2	Transport vehicle; as Specification Clause 184.9 for use after issue of Substantial Completion Certificate.	wk	20			
A229	Set of progress photographs comprising six prints.	nr	200			
	Equipment for use by the Engineer's staff.					
A231.1	Establishment and removal of office equipment.	sum				
A231.2	Maintenance of office equipment.	sum				
A231.3	Maintenance of office equipment after issue of Substantial Completion Certificate.	wk	20			
A232.1	Establishment and removal of laboratory equipment.	sum				
A232.2	Maintenance of laboratory equipment.	sum				
A233.1	Establishment and removal of surveying equipment.	sum				
A233.2	Maintenance of surveying equipment.	sum				
				PAGE TOTAL		

Number	Item description	Unit	Quantity	Rate	Amount £	Amount p
	Specified requirements.					
A242	Attendance upon the Engineer's staff, chainmen.	wk	104			
A243	Attendance upon the Engineer's staff, laboratory assistants.	wk	104			
A250	Testing of materials; concrete test cubes; samples and methods of testing as Specification Clauses 186.1 to 186.9.	nr	400			
	Testing of the Works.					
A260.1	Clay pipes normal bore not exceeding 200 mm, test as Specification Clause 187.1.	sum				
A260.2	Clay pipes nominal bore 200–300 mm, test as Specification Clause 187.2.	sum				
A260.3	Spun iron pipes nominal bore not exceeding 200 mm, test as Specification Clause 187.3.	sum				
A260.4	Spun iron pipe nominal bore 200–300 mm, test as Specification Clause 187.3.	sum				
A260.5	Aeration tanks; watertightness test as Specification Clause 188.1.	sum				
A260.6	Final settling tanks; watertightness test as Specification Clause 189.1.	sum				
A260.7	Storm tanks; watertightness test as Specification Clause 190.1.	sum				
	Temporary Works.					
A271.1	Traffic diversions.	sum				
A272.1	Traffic regulation; establishment and removal.	sum				
A272.2	Traffic regulation; continuing operation and maintenance.	wk	104			
A273.1	Establishment and removal of access roads.	sum				
A273.2	Maintenance of access roads.	sum				
A276.1	Establishment and removal of pumping plant.	sum				
A273.2	Maintenance of access roads.	wk	104			
A276.1	Establishment and removal of pumping plant.	sum				
A276.2	Operation and maintenance of pumping plant.	h	500			
A277.1	Establishment and removal of de-watering plant.	sum				
A277.2	Operation and maintenance of de-watering plant.	wk	17			
				PAGE TOTAL		

Number	Item description	Unit	Quantity	Rate	Amount	
					£	p
	Method-Related Charges.					
	Accommodation and buildings.					
A311.1	Set up offices; Fixed.	sum				
A311.2	Maintain offices for the duration of construction; Time-Related.	sum				
A311.3	Remove offices; Fixed.	sum				
A314.1	Set up stores and materials compound; Fixed.	sum				
A314.2	Remove stores and materials compound; Fixed.	sum				
A315.1	Set up canteens and messrooms; Fixed.	sum				
A315.2	Operate canteens and messrooms for the duration of construction; Time-Related.	sum				
	Services.					
A321	Set up electricity supply and standby generator; Fixed.	sum				
A322	Water supply for the duration of construction; Time-Related.	sum				
A325	Site transport for the duration of construction comprising one 5 t truck and two tractors and trailers; Time-Related.	sum				
A327	Welfare facilities complying with HSAW regulations for the duration of construction; Time-Related.	sum				
	Plant.					
	35 t crane for excavation and concreting of main tanks.					
A331.1	Bring to Site; Fixed.	sum				
A331.2	Operate; Time-Related.	sum				
A331.3	Remove; Fixed.	sum				

Completed by tenderer PAGE TOTAL

Number	Item description	Unit	Quantity	Rate	Amount £	Amount p
	Team comprising two pusher/ripper bulldozers, six motor scrapers and one spreader bulldozer for main embankment shale filling.					
A333.1	Bring to Site twice and remove twice; Fixed.	sum				
A333.2	Operate; Time-Related.	sum				
A339.1	Bring to Site and remove jacks and other prestressing plant for stressing bridge 12 deck beams; Fixed.	sum				
A339.2	Bring to Site and remove 3 grout pans and 2 grout pumps for work in items 2.B111.1 to 2.B345 inclusive; Fixed.	sum				
	Temporary Works.					
	Road diversion at Newton Street during construction culvert 2 lb.					
A351.1	Install; Fixed.	sum				
A351.2	Operate and maintain; Time-Related.	sum				
A351.3	Remove; Fixed.	sum				
A353	Install access road, Site entrance to batching plant near downstream tunnel portal; Fixed.	sum				
A356	Pumping as required during excavation and concreting of tower foundations; Time-Related.	sum				
A361.1	Erect scaffolding surrounding administration block; Fixed.	sum				
A361.2	Hire of scaffolding surrounding administration block from completion of frame until completion of brickwork; Time-Related.	sum				
	Erection of falsework to support bridge deck formwork; Fixed.					
A362.1	Bridge 1.	sum				
A362.2	Bridge 2.	sum				
A362.3	Bridge 3.	sum				
A362.4	Bridge 4.	sum				
Completed by tenderer				PAGE TOTAL		

Number	Item description	Unit	Quantity	Rate	Amount	
					£	p
	Method-Related Charges.					
	Temporary Works.					
A363	Temporary sheet pile wall to retain excavation on South side of Long Lane diversion from chainage 125 to 275; Fixed.	sum				
A364.1	Make 4 sets of full height forms for treated water reservoir columns; Fixed.	sum				
A364.2	Make 120 raked side panels and 20 rectangular end panels for formwork to 3 m high lifts in main dam concreting; Fixed.	sum				
A364.3	Use 15 m long circular steel form for lining diversion tunnel; Time-Related.	sum				
	Supervision and labour					
A371.1	Management and supervision for the duration of construction; Time-Related.	sum				
A371.2	Additional management and supervision during construction of the main pumphouse; Time-Related.	sum				
A372	Administration for the duration of construction; Time-Related.	sum				
A373.1	Labour for maintenance of plant and site services during earth moving and concreting operations; Time-Related.	sum				
A373.2	Labour for concreting gang during concreting operations at main treatment works site; Time-Related.	sum				
Completed by tenderer				PAGE TOTAL		

Number	Item description	Unit	Quantity	Rate	Amount £	p
	Provisional Sums.					
	Daywork.					
A411	Labour.	sum			50 000	00
A412	Percentage adjustment to Provisional Sum for Daywork labour.	%				
A413	Materials.	sum			25 000	00
A414	Percentage adjustment to Provisional Sum for Daywork materials.					
A415	Plant.					
A416	Percentage adjustment to Provisional Sum for Daywork plant.	%				
A417	Supplementary charges.	sum			20 000	00
A418	Percentage adjustment to Provisional Sum for Daywork supplementary charges.	%				
	Other Provisional Sums.					
A420.1	Permanent diversion of existing services.	sum			20 000	00
A420.2	Landscaping.	sum			25 000	00
A420.3	Repairs to existing tanks.	sum			15 000	00
A420.4	Repairs to existing pipelines.	sum			10 000	00
				PAGE TOTAL		

Number	Item description	Unit	Quantity	Rate	Amount	
					£	p
	Nominated Sub-contracts which Include work on the Site.					
A510.1	Electrical installation.	sum			50 000	00
A520.1	Labours.	sum				
A540.1	Other charges and profit.	%				
A510.2	Flow recording and control equipment.	sum			25 000	00
A520.2	Labours.	sum				
A540.2	Flow charges and profit.	%				
A510.3	Pumping machinery.	sum			42 000	00
A530	Special labours; attendance on commissioning and testing as Specification Clause 207.1, carried out upon completion of pump installation.	sum				
A540.3	Other charges and profit.	%				
	Nominated Sub-contracts which do not include work on the Site.					
A610.1	Precast concrete bridge beams.	sum			37 000	00
A620.1	Labours.	sum				
A640.1	Other charges and profit.	%				
A610.2	Precast concrete filter wall units.	sum			12 000	00
A620.2	Labours.	sum				
A640.2	Other charges and profit.	%				
				PAGE TOTAL		

Number	Item description	Unit	Quantity	Rate	Amount £	Amount p
	GROUND INVESTIGATION.					
	TRIAL PITS AND TRENCHES.					
B113	Number in material other than rock maximum depth 2–3 m; minimum plan area at the bottom of the excavation 2 m^2.	nr	6			
B130	Depth in material other than rock; minimum plan area at the bottom of the excavation 2 m^2.	m	18			
B150	Depth supported.	m	18			
B160	Depth backfilled with excavated material.	m	18			
B170	Removal of obstructions.	h	5			
B180	Pumping at a minimum extraction rate of 7000 litres per hour.	h	12			
	LIGHT CABLE PERCUSSION BOREHOLES.					
	Nominal diameter at base 150 mm.					
B210	Number.	nr	8			
B231	Depth in holes of maximum depth not exceeding 5 m.	m	20			
B232	Depth in holes of maximum depth 5–10 m.	m	10			
B233	Depth in holes of maximum depth 10–20 m.	m	15			
B234	Depth in holes of maximum depth 20–30 m.	m	50			
B260	Depth backfilled with excavated material.	m	95			
B270	Chiselling to prove rock or to penetrate obstructions.	h	10			
				PAGE TOTAL		

Number	Item description	Unit	Quantity	Rate	Amount £	Amount p
	ROTARY DRILLED BOREHOLES.					
	Minimum core diameter 75 mm.					
B310	Number; continuations of light cable percussion boreholes.	nr	2			
B343	Depth with core recovery in holes of maximum depth 10–20 m.	m	30			
B360	Depth backfilled with cement grout.	m	10			
B370	Core boxes, core 3.0 m long.	nr	4			
	SAMPLES.					
B412	Disturbed samples of soft material from the surface or from trial pits and trenches: minimum 5 kg; Class 3.	nr	18			
B421	Open tube samples from boreholes. 100 mm diameter × 450 mm long, undisturbed sample; Class 1.	nr	40			
B422	Disturbed samples from boreholes: minimum 5 kg, Class 3.	nr	25			
B423	Groundwater samples from boreholes: minimum 1 litre.	nr	8			
	SITE TESTS AND OBSERVATIONS.					
B512	Groundwater level; taken at commencement and end of work each day.	nr	8			
B513	Standard penetration test; in light cable percussion boreholes.	nr	25			
B515	Vane test in borehole.	nr	5			
	INSTRUMENTAL OBSERVATIONS.					
	Pressure head: weekly observations, protected with chestnut fencing.					
B611	Standpipes.	m	18			
B613	Install covers.	nr	3			
B614	Standpipe readings.	nr	9			
				PAGE TOTAL		

Number	Item description	Unit	Quantity	Rate	Amount £	Amount p
	LABORATORY TESTS.					
	Classification.					
B711	Moisture content.	nr	10			
B712	Atterberg limits.	nr	20			
B714	Particle size analysis by sieve.					
B715	Particle size analysis by pipette or hydrometer.	nr	4			
	Chemical content.					
B722	Sulphate.	nr	16			
B723	pH value.	nr	8			
	Consolidation.					
B741	Oedometer cell.	nr	6			
	Soil strength.					
B761	Quick undrained triaxial set of three 38 mm diameter specimens.	nr	20			
	Rock strength.					
B775	Point load tests on core samples as Specification Clause B243.	nr	6			
	PROFESSIONAL SERVICES.					
B831	Engineer or geologist graduate.	h	5			
B832	Engineer or geologist chartered.	h	15			
B833	Engineer or geologist principal or consultant.	h	5			
B840	Visits to the Site.	nr	2			
				PAGE TOTAL		

C

Number	Item description	Unit	Quantity	Rate	Amount	
					£	p
	GEOTECHNICAL AND OTHER SPECIALIST PROCESSES.					
	Drilling for grout holes.					
	Diameter 50 mm.					
	Vertically downwards.					
C111	In holes of depth: not exceeding 5 m.	m	80			
C112	In holes of depth: 5–10 m.	m	160			
C113	In holes of depth: 20–30 m.	m	310			
C114	In holes of depth: 20–30 m.	m	630			
C116	In holes of depth: 50 m.	m	200			
	Downwards at an angle 0°–45° to the vertical.					
C123	In holes of maximum depth: 10–20 m.	m	620			
C124	In holes of maximum depth: 20–30 m.	m	3300			
	Grout holes diameter 50 mm.					
C410	Number of holes.	nr	234			
C420	Number of stages.	nr	702			
C430	Single water pressure tests.	nr	20			
	Grout materials and injection.					
	Materials.					
C511	Cement; to BS 12.	t	154			
C512	Pulverised fuel ash; as Specification Clause G420.7.	t	290			
C513	Sand; to BS 1199.	t	50			
	Injection.					
C521.1	Number of injections; in descending stages.	nr	50			
C521.2	Number of injections; in ascending stages.	nr	75			
C522	Neat cement grout.	t	25			
C523.1	Cement and sand grout.	t	200			
C523.2	Cement and pulverised fuel ash grout.	t	269			
C526	Single packer settings.	nr	50			
				PAGE TOTAL		

Number	Item description	Unit	Quantity	Rate	Amount £	Amount p
	Diaphragm wall for anchor chamber as drawing 134/65 thickness 1.05 m.					
C613	Excavation maximum depth 10–15 m.	m^3	2400			
C623	Excavation in rock maximum depth 10–15 m.	m^3	400			
C640	Concrete; with sulphate resisting cement strength 30 N/mm^2.	m^3	2730			
C664	High yield steel bar reinforcement to BS 4449 diameter 12 mm.	t	40			
C665	High yield steel bar reinforcement to BS 4449 diameter 16 mm.	t	66			
C666	High yield steel bar reinforcement to BS 4449 diameter 20 mm.	t	120			
C670	Waterproofed joints; as Specification Clause C47.	sum				
C680	Guide walls.	m	242			
	Temporary ground anchorages: East Quay wall; horizontal load 27 tonnes/metre; water and grout testing, pregrouting and grouting as Specification Clause W7/27.					
C711	Number maximum depth 25 m.	nr	7			
C721	Total length of tendons.	m	140			
	Sand drains.					
C811	Number of drains across-sectional dimension not exceeding 100 m.	nr	50			
C851	Depth of drains of maximum depth: 10–15 m cross-sectional dimension not exceeding 100 mm; type C2 granular material.	m	650			
				PAGE TOTAL		

C

Number	Item description	Unit	Quantity	Rate	Amount £	Amount p
	PART 5. DEMOLITION AND SITE CLEARANCE.					
D100	General clearance.	ha	27			
D220	Trees girth 1–2 m; holes backfilled with excavated material.	nr	39			
D250	Trees girth exceeding 5 m.	nr	30			
D320	Stumps diameter 500 mm–1 m.	nr	21			
D414	Buildings brickwork volume 250–500 m^3; farmhouse at Farley Court.	sum				
D435	Buildings masonry volume 500–1000 m^3; barn at Farley Court.	sum				
D545	Other structures metal volume 500–1000 m^3; screening plant to remain the property of the Employer.	sum				
D620	Pipelines nominal bore 300–500 mm.	m	110			
D900	Pump removed, cleaned and delivered to Employer's depot at Church Stretton.	nr	2			
				PAGE TOTAL		

Number	Item description	Unit	Quantity	Rate	Amount	
					£	p
	EARTHWORKS.					
E120	Excavation by dredging; Mersey estuary adjacent to Sandon Dock.	m^3	10 971			
E311	Excavation for foundations topsoil maximum depth: not exceeding 0.25 m.	m^3	37			
E324	Excavation for foundations maximum depth: 1–2 m.	m^3	170			
E324.1	Excavation for foundations maximum depth: 1–2 m; around pile shafts.	m^3	23			
E324.2	Hand excavation for foundations maximum depth: 1–2 m; Commencing Surface underside of blinding of outfall.	m^3	15			
	General excavation.					
E415	Topsoil maximum depth 2–5 m.	m^3	153			
E425	Maximum depth 2–5 m.	m^3	17 350			
E435	Rock maximum depth: 2–5 m.	m^3	750			
E444	Mass concrete exposed at the Commencing Surface maximum depth: 1–2 m.	m^3	53			
E454	Reinforced concrete not exposed at the Commencing Surface maximum depth: 1–2 m.	m^3	112			
	Excavation ancilliaries.					
E512.1	Trimming of excavated surfaces.	m^2	7500			
E512.2	Trimming of surfaces; excavated by dredging.	m^2	5271			
E552	Preparation of excavated surfaces.	m^2	15 763			
				PAGE TOTAL		

E

Number	Item description	Unit	Quantity	Rate	Amount	
					£	p
	EARTHWORKS.					
	Excavation ancillaries.					
E523	Preparation of excavated rock surfaces.	m^2	576			
E524	Preparation of excavated surfaces; mass concrete.	M^2	110			
E532.1	Disposal of excavated material.	m^3	2700			
E532.2	Disposal of material excavated by dredging.	m^3	10 971			
E542	Double handling of excavated material.	m^3	10 350			
E560	Excavation of material below the Final Surface and replacement with grade B granular material.	m^3	1000			
E570	Timber supports left in.	m^2	300			
E580	Metal supports left in.	m^2	500			
E613	Filling to structures.	m^3	527			
E624	Filling to embankments, selected excavated material other than topsoil or rock.	m^3	100 716			
E712	Filling ancillaries trimming of filled surfaces; inclined at an angle of 10°–45° to the horizontal.	m^2	1054			
E722	Filling ancillaries preparation of filled surfaces.	m^2	357			
				PAGE TOTAL		

Number	Item description	Unit	Quantity	Rate	Amount	
					£	p
	PART 27. ROADS.					
	EARTHWORKS.					
E210	Excavation for cuttings; topsoil.	m^3	584			
E220.1	Excavation for cuttings; Excavated Surface 0.25 m above Final Surface.	m^3	10 587			
E220.2	Excavation for cuttings; Commencing Surface 0.25 m above Final Surface.	m^3	680			
E522	Excavation ancillaries preparation of excavation surface.	m^2	5110			
E532	Excavation ancillaries disposal of excavated material; to on Site spoil heap at location A on drawing 7/47.	m^3	584			
E624	Filling embankments selected excavated material other than topsoil or rock.	m^3	5834			
E712	Filling ancillaries preparation of filled surfaces.	m^2	2787			
E730	Filling ancillaries geotextiles; melded fibre mat grade G2 laid upon surfaces inclined at an angle of 10°–45° to the horizontal.	m^2	1250			
				PAGE TOTAL		

E

Number	Item description	Unit	Quantity	Rate	Amount	
					£	p
	PART 28. SITEWORKS.					
	EARTHWORKS.					
E411	General excavation topsoil maximum depth: not exceeding 0.25 m.	m^3	3482			
E424	General excavation maximum depth: 1–2 m.	m^3	25 234			
E532	Disposal of excavated material.	m^3	9924			
E624	Filling embankments selected excavated material other than topsoil or rock.	m^3	15 310			
E641	Filling thickness 150 mm excavated topsoil.	m^2	23 213			
E642	Filling thickness 150 mm imported topsoil.	m^2	1573			
E711	Filling ancillaries trimming of filled surfaces, topsoil.	m^2	17 036			
E810	Landscaping turfing.	m^2	5700			
E830	Landscaping grass seeding upon a surface inclined at an angle to the horizontal exceeding 10°.	m^2	2050			
E860.1	Landscaping oak trees exceeding 5 m high.	nr	110			
E860.2	Landscaping sycamore trees exceeding 5 m high.	nr	120			
				PAGE TOTAL		

Number	Item description	Unit	Quantity	Rate	Amount £	Amount p
	IN SITU CONCRETE.					
	Provision of concrete.					
F143	Standard mix ST4 cement to BS 12, 20 mm aggregate to BS 882.	m^3	840			
	Placing of concrete.					
	Reinforced.					
F634	Suspended slab thickness: exceeding 500 mm; voided bridge deck.	m^3	462			
F654	Columns and piers cross-sectional area: 0.25–1 m^2.	m^3	96			
F664	Beams cross-sectional area: 0.25–1 m^2.	m^3	38			
F665	Beams cross-sectional area: exceeding 1 m^2.	m^3	53			
F666.1	Beams special sections: type J1 drawing 137/26.	m^3	87			
F666.2	Beams special sections: type J2 drawing 137/26.	m^3	104			
				PAGE TOTAL		

F

Number	Item description	Unit	Quantity	Rate	Amount £	Amount p
	IN SITU CONCRETE.					
	Provision of concrete, standard mix.					
F133	ST3 cement to BS 12, 20 mm aggregate to BS 882.	m^3	350			
F144	ST4 cement to BS 12, 40 mm aggregate to BS 882.	m^3	1032			
F148	ST4 sulphate resisting cement to BS 4027, 40 mm aggregate to BS 382.	m^3	504			
F253	Provision of concrete designed mix for ornamental surfaces grade C25 cement to BS 12, 20 mm aggregate; minimum cement content 250 kg/m^3 sampling as Specification Clause 252.	m^3	632			
	Placing of concrete.					
F511	Mass blinding thickness not exceeding 150 mm.	m^3	97			
F580.1	Mass backfilling around structure; beneath inlet channel placed against an excavated surface.	m^3	210			
F580.2	Mass benching; wet wells.	m^3	21			
F580.3	Mass filling to flume; inlet channel to pumping station.	m^3	5			
F580.4	Mass plinths; 900 × 900 × 1000 mm.	m^3	7			
F580.5	Mass plinths; 1500 × 1500 × 1500 mm.	m^3	10			
	Reinforced.					
F623.1	Bases and ground slabs thickness: 300–500 mm.	m^3	906			
F623.2	Ground slabs thickness: 300–500 mm; to be placed in one continuous pour in filter block base.	m^3	15			
F633	Suspended slabs thickness: 300–500 mm.	m^3	120			
F641.1	Walls thickness: not exceeding 150 mm.	m^3	31			
F641.2	Walls thickness: not exceeding 150 mm; weir wall top finished to precise line and level.	m^3	10			
F642	Walls thickness: 150–300 mm.	m^3	290			
F643	Walls thickness: 300–500 mm.	m^3	520			
F653	Columns and piers cross-sectional area: 0.1–0.25 m^2.	m^3	97			
				PAGE TOTAL		

Number	Item description	Unit	Quantity	Rate	Amount	
					£	p
	IN SITU CONCRETE.					
	Placing of concrete.					
	Reinforced.					
F654	Coldmns and piers cross-sectional area: 0.25–1 m^2.	m^3	94			
F655	Columns and piers cross-sectional area: exceeding 1 m^2; column cap.	m^3	10			
F664	Beams cross-sectional area: 0.25–1 m^2.	m^3	70			
F666.1	Beams special sections; 600 × 900 mm (internal) box beam, wall thickness: 150 mm.	m^3	67			
F666.2	Beams special sections; type J3 drawing 137/27.	m^3	76			
F666.3	Beams special sections; type J4 drawing 137/27.	m^3	98			
F666.4	Beams special sections; type J5 drawing 137/27.	m^3	66			
	Other concrete forms.					
F680.1	Diffuser drum thickness 150 mm, diffuser type D1 drawing 137/49.	m^3	3			
F680.2	Box culvert internal dimensions 1.5 × 3 m wall thickness: 250 mm.	m^3	78			
F680.3	Box culvert internal dimensions 2 × 4 m wall thickness: 250 mm.	m^3	104			
F680.4	Steps and stairs thickness: 100–300 mm.	m^3	12			
				PAGE TOTAL		

F

Number	Item description	Unit	Quantity	Rate	Amount	
					£	p
	CONCRETE ANCILLARIES.					
G112	Formwork rough finish horizontal width 0.1–0.2 m.	m	129			
G113	Formwork rough finish horizontal width 0.2–0.4 m to be left in.	m^2	7			
G145	Formwork rough finish vertical.	m^2	370			
G155	Formwork rough finish curved to one radius in one plane; radius 10.5 m, back of wall D drawing 137/6.	m^2	997			
G184	Formwork rough finish for concrete component of constant cross-section; box culvert internal dimensions 1.5 × 3 m wall thickness: 250 mm.	m	326			
	Formwork fair finish.					
G211	Horizontal width not exceeding 0.1 m.	m	410			
G213	Horizontal width 0.2–0.4 m.	m^2	331			
G215	Horizontal.	m^2	981			
G222.1	Sloping width 0.1–0.2 m.	m	723			
G222.2	Sloping width 0.1–0.2 m; upper surface.	m	27			
G225.1	Sloping.	m^2	128			
G225.2	Sloping; upper surface.	m^2	157			
G235	Battered.	m^2	429			
G241	Vertical width not exceeding 0.1 m.	m	410			
G242	Vertical width 0.1–0.2 m.	m	924			
G243	Vertical width 0.2–0.4 m.	m^2	192			
G245	Vertical.	m^2	780			
G252	Curved to one radius in one plane width 0.15 m; radius 1.0 m, column plinths.	m	927			
G255	Curved to one radius in one plane radius 11.5 m; face of wall D drawing 137/6.	m^2	1354			
G260.1	Spherical radius 2 m; flume chamber.	m^2	43			
G260.2	Conical maximum radius 600 mm minimum radius 300 mm; outlet chamber.	m^2	27			
G260.3	Conical maximum radius 800 mm minimum radius 400 mm; inlet chamber.	m^2	35			
				PAGE TOTAL		

Number	Item description	Unit	Quantity	Rate	Amount £	Amount p
	CONCRETE ANCILLARIES.					
	Formwork fair finish.					
G271	For small voids depth not exceeding 0.5 m.	nr	157			
G272	For small voids depth 0.5–1 m.	nr	298			
G276	For large voids depth 2.3 m.	nr	76			
G278.1	For large voids depth 2.3 m.	nr	5			
G278.2	For large voids depth 2.6 m.	nr	1			
	For concrete components of constant cross-section.					
G281.1	Beams; 200 × 300 mm beam B27.	m	37			
G281.2	Beams; 300 × 400 mm beam B28.	m	50			
G281.3	Beams; 450 × 600 mm beam B30.	m	46			
G282.1	Columns; 450 × 450 mm columns 1–24.	m	98			
G282.2	Columns; 600 × 600 mm columns 25–30.	m	27			
G282.3	Columns; curved to 150 mm radius columns 31–54.	m	95			
G284	Box culvert internal dimension 1.5 × 3 m wall thickness 250 mm beneath entrance road.	m	310			
G285	Projections.	m	2076			
G286	Intrusions.	m	1594			
G299	For diffuser drum including perforations as detailed on drawing 137/54.	m	5			
				PAGE TOTAL		

G

G

Number	Item description	Unit	Quantity	Rate	Amount	
					£	p
	CONCRETE ANCILLARIES.					
	Reinforcement.					
	Plain round steel bars to BS 4449.					
G511	Diameter 6 mm.	t	3.7			
G512	Diameter 8 mm.	t	3.9			
G513	Diameter 10 mm.	t	4.2			
G514	Diameter 12 mm.	t	4.9			
G515.1	Diameter 16 mm.	t	5.7			
G515.2	Diameter 16 mm; length 15 m.	t	2.1			
G516.1	Diameter 20 mm.	t	5.9			
G516.2	Diameter 20 mm; length 15 m.	t	1.0			
G516.3	Diameter 20 mm; length 18 m.	t	0.9			
G517.1	Diameter 25 mm.	t	6.3			
G517.2	Diameter 25 mm; length 15 m.	t	1.2			
G518	Diameter 32 mm or greater.	t	7.6			
G526.1	Deformed high yield steel bars to BS 4449 diameter 20 mm.	t	4.2			
G526.2	Deformed high yield steel bars to BS 4449 diameter 20 mm; length 15 m.	t	2.1			
G528	Deformed high yield steel bars to BS 4449 diameter 32 mm or greater.	t	3.9			
G563.1	Steel fabric to BS 4483 nominal mass 3–4 kg/m^2; reference A193.	m^2	3728			
G563.2	Steel fabric to BS 4483 nominal mass 3–4 kg/m^2; reference A252.	m^2	1807			
G564	Steel fabric to BS 4483 nominal mass 4–5 kg/m^2; reference C503.	m^2	286			
				PAGE TOTAL		

Number	Item description	Unit	Quantity	Rate	Amount £	Amount p
	CONCRETE ANCILLARIES.					
	Joints.					
	25 mm bitumen impregnated fibreboard filler.					
G621	Open surface width: not exceeding 0.5 m.	m^2	97			
G622	Open surface width: 0.5–1 m.	m^2	38			
G641	Formed surface width: not exceeding 0.5 m.	m^2	107			
G642	Formed surface width: 0.5–1 m.	m^2	98			
G643.1	Formed surface width: 1.2 m.	m^2	27			
G643.2	Formed surface width: 1.45 m.	m^2	41			
G651	PVC centre bulb waterstop; 100 mm wide.	m	38			
G652	PVC centre bulb waterstop; 175 mm wide.	m	110			
G653	Rubber centre bulb waterstop; 225 mm wide.	m	150			
G654.1	Rubber centre bulb waterstop; 400 mm wide.	m	28			
G654.2	Rubber centre bulb waterstop; 450 mm wide.	m	64			
G670.1	Rebate sealed with 25 × 25 mm Plastic or similar approved.	m	125			
G670.2	Rebate sealed with 25 × 25 mm Plastijoint or similar approved.	m	670			
G682.1	Sleeved and capped dowels; 25 mm diameter by 300 mm mild steel bars at 0.5 m centres as detail E drawing 137/25.	nr	128			
G682.2	Sleeved and capped dowels; 25 mm diameter by 300 mm mild steel bars at 1 m centres as detail E drawing 137/25.	nr	31			
				PAGE TOTAL		

G

Number	Item description	Unit	Quantity	Rate	Amount	
					£	p
	CONCRETE ANCILLARIES.					
	Concrete accessories.					
G811	Finishing of top surfaces wood float.	m^2	4976			
G812	Finishing of top surfaces steel trowel.	m^2	10 821			
G814.1	Finishing of top surfaces granolithic finish; as Specification Clause G8/05, 40 mm thick; steel trowel surface treatment.	m^2	192			
G814.2	Finishing of top surfaces granolithic finish; as Specification Clause G8/05, 55 mm thick; steel trowel surface treatment.	m^2	141			
G815	Finishing of top surfaces cement and sand screed as Specification Clause G8/12, 40 mm thick; steel trowel surface treatment.	m^2	98			
G822	Finishing of formed surfaces bush hammering.	m^2	487			
	Inserts.					
G831	100 mm diameter GVC pipe; excluding supply of the pipe.	m	27			
G832.1	10 mm diameter, 25 mm deep expanding bolt projecting from one surface.	nr	110			
G832.2	15 mm diameter, 40 mm deep rag bolt projecting from one surface.	nr	110			
G832.3	25 mm diameter, 100 mm long mild steel holding down bolt projecting from one concrete surface for scraper machinery as detail L drawing 138/17.	nr	64			
G832.4	100 mm diameter ductile iron pipe projecting from one surface; excluding supply of pipe.	nr	10			
G832.5	150 mm diameter ductile iron pipe with puddle flange projecting from two surfaces; excluding supply of pipe.	nr	3			
G832.6	225 mm diameter ductile iron pipe projecting from two surfaces grouted with grout type G3 into 450 × 450 mm preformed opening; excluding supply and fixing of the pipe under separate contract.	nr	4			
G842	Grouting under plates area 0.1–0.5 m^2 with type G2 grout.	nr	8			
				PAGE TOTAL		

Number	Item description	Unit	Quantity	Rate	Amount £	p
	PRECAST CONCRETE.					
	Ordinary prescribed mix concrete grade C40 as Specification Clause F2/36.					
	Strefford Bridge Deck.					
H113.1	Secondary beams inverted tee 180 × 400 mm length 4.25 m mass 500 kg-lt mark B1.	nr	32			
H113.2	Secondary end beams rectangular 200 × 450 mm length 4.25 m mass 500 kg-lt mark B2.	nr	4			
H268	Prestressed pre-tensioned main beams I section 800 × 1400 mm length 23.1 m mass 31 t mark B3 prestressing as drawing 136/21 and Specification Clauses H2/21–27.	nr	2			
H523	Service duct cover slabs thickness 100 mm area 1–4 m^2 mass 500 kg-lt mark S1.	nr	28			
H810	Parapet coping units filleted rectangular 250 × 220 mm cross-sectional area 0.05 m^2 mass 120 kg/m mark C1.	m	124			
				PAGE TOTAL		

Number	Item description	Unit	Quantity	Rate	Amount	
					£	p
	PIPEWORK – PIPES.					
	Clay pipes to BS 65 with spigot and socket flexible joints nominal bore 225 mm in trenches.					
	Between manholes 7 and 11.					
I123.1	Depth: 1.5–2 m.	m	187			
I124.1	Depth: 2–2.5 m.	m	291			
I125	Depth: 2.5–3 m.	m	102			
	Between manholes 27 and 31.					
I123.2	Depth: 1.5–2 m.	m	113			
I124.2	Depth: 2–2.5 m.	m	202			
	Prestressed concrete pipes to BS 5911 (class M) with ogee joints nominal bore 375 mm.					
I231	Between manholes 18 and 19 installed by thrust boring under Central Wales line railway embankment.	m	98			
	Between manholes 19 and 23 in trenches. Commencing Surface underside of topsoil. (Stripping measured separately in item number K760.1).					
I234	Depth: 2–2.5 m.	m	127			
I235	Depth: 2.5–3 m.	m	78			
I238	Depth: 4.5–5 m.	m	21			
				PAGE TOTAL		

Number	Item description	Unit	Quantity	Rate	Amount £	Amount p
	PIPEWORK – PIPES.					
	Prestressed concrete pipes to BS 5911 (class M) with ogee joints nominal bore 375 mm.					
	Manhole 23 to primary tank.					
I233	Depth 1.5–2 m; in trench laid alongside second pipe.	m	27			
	Ductile spun iron pipes to BS 4622 (class 3) with Forsoam joints nominal bore 700 mm.					
I341	Feed pipe supported above the ground.	m	16			
I343	In trench depth 1.5–2 m Commencing Surface underside of main tank slab; draw off pipe main tank collector to valve chamber.	m	12			
I345	In trench D depth 2.5–3 m bypass pressure main valve chamber to circulation unit.	m	33			
I346	In trench D depth 3–3.5 m service main valve chamber to circulation unit.	m	33			
	UPVC to BS 3505 with compression joints nominal bore 50 mm washwater mains.					
	In trenches.					
I512	Depth: not exceeding 1.5 m; excavation by hand.	m	15			
I513	Depth: 1.5–2 m.	m	150			
I514	Depth: 2–3 m.	m	128			
				PAGE TOTAL		

Number	Item description	Unit	Quantity	Rate	Amount	
					£	p
	PIPEWORK – FITTINGS AND VALVES.					
	Clay pipe fittings to BS 65 with spigot and socket flexible joints: nominal bore 225 mm.					
J112	Bends.	nr	385			
J122	Junctions and branches.	nr	127			
J132	Tapers.	nr	38			
	Prestressed concrete pipe fittings to BS 5911 (class M) with ogee joints.					
J212	Bends; nominal bore 225 mm.	nr	41			
J213.1	Bends; nominal bore 375 mm.	nr	141			
J213.2	Bends; nominal bore 450 mm.	nr	19			
J214	Bends; nominal bore 825 mm.	nr	47			
J222	Junctions and branches; nominal bore 225 mm.	nr	27			
J223.1	Junctions and branches; nominal bore 375 mm.	nr	76			
J223.2	Junctions and branches; nominal bore 450 mm.	nr	8			
J224	Junctions and branches; nominal bore 825 mm.	nr	23			
J232	Tapers; nominal bore 225 mm.	nr	16			
J233	Tapers; nominal bore 450 mm.	nr	16			
J234	Tapers; nominal bore 825 mm.	nr	4			
				PAGE TOTAL		

Number	Item description	Unit	Quantity	Rate	Amount £	Amount p
	PIPEWORK – FITTING AND VALVES.					
	Ductile spun iron pipe fittings to BS 4622 (class 3) with Grippon joints.					
J311	Bends; normal bore 150 mm.	nr	27			
J313.1	Bends; nominal bore 400 mm 22.5° effective length 220 mm; not in trenches.	nr	1			
J313.2	Bends; nominal bore 400 mm 45 effective length 400 mm.	nr	6			
J313.3	Bends; nominal bore 500 mm vertical 90° effective length 840 mm.	nr	3			
J353	Adaptors; nominal bore 400 mm Grippon to flange effective length 160 mm.	nr	6			
J382.1	Straight specials nominal bore 225 mm effective length 1220 mm.	nr	1			
J382.2	Straight specials nominal bore 225 mm effective length 1500 mm; with puddle flange.	nr	1			
J383	Straight specials nominal bore 400 mm effective length 900 mm; not in trenches.	nr	3			
J393	Blank flanges; nominal bore 400 mm.	nr	3			
	Cast iron gate valves hand operated to BS 3464 (type a) with extension spindles.					
J813.1	400 mm nominal bore with T key.	nr	4			
J813.2	400 mm nominal bore with hand wheel.	nr	2			
J813.3	500 mm nominal bore with hand wheel.	nr	3			
	Cast iron valves as Holdwater Industries catalogue numbers stated or similar approved.					
J835	Double hinged flap valves nominal bore 1100 mm nr 890.	nr	2			
J884.1	Penstocks nominal bore 700 mm nr 731 with handwheel headstock and foot bracket; invert to handwheel distance 2050 mm.	nr	6			
J884.2	Penstocks nominal bore 700 mm nr 740 with T key headstock and foot bracket; invert to T key distance 3100 mm.	nr	9			
J894	Sludge draw off valves nominal bore 700 mm nr 780.	nr	3			
				PAGE TOTAL		

Number	Item description	Unit	Quantity	Rate	Amount	
					£	p
	PIPEWORK – FITTING AND VALVES.					
	Ductile spun iron pipe fittings to BS 4622 (class 3) with Grippon joints.					
J311	Bends; normal bore 150 mm.	nr	27			
J313.1	Bends; nominal bore 400 mm 22.5° effective length 220 mm; not in trenches.	nr	1			
J313.2	Bends; nominal bore 400 mm 45 effective length 400 mm.	nr	6			
J313.3	Bends; nominal bore 500 mm vertical 90° effective length 840 mm.	nr	3			
J353	Adaptors; nominal bore 400 mm Grippon to flange effective length 160 mm.	nr	6			
J382.1	Straight specials nominal bore 225 mm effective length 1220 mm.	nr	1			
J382.2	Straight specials nominal bore 225 mm effective length 1500 mm; with puddle flange.	nr	1			
J383	Straight specials nominal bore 400 mm effective length 900 mm; not in trenches.	nr	3			
J393	Blank flanges; nominal bore 400 mm.	nr	3			
	Cast iron gate valves hand operated to BS 3464 (type a) with extension spindles.					
J813.1	400 mm nominal bore with T key.	nr	4			
J813.2	400 mm nominal bore with hand wheel.	nr	2			
J813.3	500 mm nominal bore with hand wheel.	nr	3			
	Cast iron valves as Hoidwater Industries catalogue numbers stated or similar approved.					
J835	Double hinged flap valves nominal bore 1100 mm nr 890.	nr	2			
J884.1	Penstocks nominal bore 700 mm nr 731 with handwheel headstock and foot bracket; invert to handwheel distance 2050 mm.	nr	6			
J884.2	Penstocks nominal bore 700 mm nr 740 with T key headstock and foot bracket; invert to T key distance 3100 mm.	nr	9			
J894	Sludge draw off valves nominal bore 700 mm nr 780.	nr	3			
				PAGE TOTAL		

I-L

Number	Item description	Unit	Quantity	Rate	Amount	
					£	p
	PIPEWORK – MANHOLES AND PIPEWORK ANCILLARIES.					
	Manholes.					
K111	Brick depth not exceeding 1.5 m; excavated by hand; type A1 with medium duty cast iron cover to BS 497 reference MB1-55.	nr	7			
K112.1	Brick depth 1.5–2 m; type A1 with heavy duty triangular cast iron cover to BS 497 reference MA-T.	nr	7			
K112.2	Brick depth 1.5–2 m; type A2 with medium duty cast iron cover to BS 497 reference MB2-55.	nr	4			
K122.1	Brick with backdrop depth 1.5–2 m; type A10 with cast iron cover to BS 497 reference MC2-60/45.	nr	5			
K122.2	Brick with backdrop depth 1.5–2 m; type A20 with cast iron cover to BS 497 reference MC1-60/60.	nr	3			
K152	Precast concrete depth 1.5–2 m; type C1 with triangular cast iron cover to BS 497 reference MA-T.	nr	33			
K211	Catch pits brick depth not exceeding 1.5 m; type K1 with light duty cast iron cover to BS 497 reference C6-24/24.	nr	6			
K360	Gullies precast concrete trapped; with medium duty straight bar gulley grating to BS 497 reference E12-12.	nr	15			
K410	Filling of French and rubble drains with graded material; type C3 granular material.	m^3	310			
K433	Trenches for unpiped rubble drains cross-sectional area 0.5–0.75 m^2.	m	54			
K434	Trenches for unpiped rubble drains cross-sectional area 0.75–1 m^2.	m	89			
K453	Rectangular section ditches cross-sectional area 0.5–0.75 m^2; lined with 1000 gauge melded fibre mat.	m	28			
				PAGE TOTAL		

Number	Item description	Unit	Quantity	Rate	Amount	
					£	p
	PIPEWORK – MANHOLES AND PIPEWORK ANCILLARIES.					
	Ducts and metal culverts.					
K542	Unglazed vitrified clay 4 way 100 mm nominal bore cable duct loose jointed in trenches depth not exceeding 1.5 m beneath roads.	m	14			
K543	Unglazed vitrified clay 4 way 100 mm nominal bore cable duct loose jointed in trenches depth 1.5–2 m beneath roads.	m	25			
K553	Sectional corrugated metal culverts as Specification Clause K56/3 nominal internal diameter 700 mm in trenches depth 1.5–2 m in outfall.	m	63			
	Crossings.					
K623.1	River width 3–10 m pipe bore 900–1800 mm.	nr	2			
K623.2	Canal width 3–10 m; pipe bore 300–900 mm.	nr	4			
K641	Hedge; pipe bore not exceeding 300 mm.	nr	30			
K671	Sewer or drain; pipe bore not exceeding 300 mm.	nr	15			
K672	Sewer or drain; pipe bore 300–900 mm.	nr	7			
K682.1	Gas main; pipe bore 300–900 mm.	nr	2			
K682.2	Existing sludge draw off pipe line: pipe bore 300–900 mm.	nr	2			
K683	Underground high voltage electric cable: pipe bore 900–1800 mm.	nr	1			
	Reinstatement.					
K711	Breaking up and temporary reinstatement of roads flexible road construction maximum depth 75 mm with 250 mm sub-base; pipe bore not exceeding 300 mm.	m	387			
K712.1	Breaking up and temporary reinstatement of roads reinforced concrete slab depth 200 mm with 250 mm type 1 sub-base; pipe bore 300–900 mm.	m	284			
				PAGE TOTAL		

Number	Item description	Unit	Quantity	Rate	Amount	
					£	p
	PIPEWORK – MANHOLES AND PIPEWORK ANCILLARIES.					
K712.2	Breaking up and temporary reinstatement of roads flexible road construction maximum depth 75 mm with 250 mm type 1 sub-base; pipe bore 300–900 mm.	m	783			
K741	Breaking up temporary and permanent reinstatement of footpaths precast concrete paving flags maximum depth 75 mm with 50 mm sand bed; pipe bore not exceeding 300 mm.	m	54			
K760	Strip topsoil from easement and reinstate in cultivated land; minimum width 20 m.	m	3726			
	Other pipework ancillaries.					
K810	Reinstatement of field drains.	m	247			
K820	Marker posts; 100 × 100 × 1500 mm hardwood set in concrete base as drawing 137/15 detail C.	nr	47			
K830	Timber supports left in excavations.	m^2	450			
K852	Connection of pipes to existing manholes; nominal bore 200–300 mm; foul water sewer including breaking into manhole, reforming benching and dealing with flows in accordance with detail 7 drawing C3/21.	nr	1			
				PAGE TOTAL		

I-L

Number	Item description	Unit	Quantity	Rate	Amount £	p
	PIPEWORK – SUPPORTS AND PROTECTION ANCILLARIES TO LAYING AND EXCAVATION.					
	Extras to excavation and backfilling in pipe trenches.					
L111	Excavation of rock.	m^3	460			
L112	Excavation of mass concrete.	m^3	65			
L113	Excavation of reinforced concrete.	m^3	85			
L115	Backfilling above the Final Surface with concrete; mix ST1.	m^3	2			
L116	Backfilling above the Final Surface with imported natural material other than rock or topsoil.	m^3	310			
L117	Excavation of natural material below the Final Surface and backfilling with concrete: mix ST1.	m^3	260			
L118.1	Excavation of natural material below the Final Surface and backfilling with type C2 granular material.	m^3	270			
L118.2	Hand excavation of natural material below the Final Surface and backfilling with type C3 granular material.	m^3	230			
	Extras to excavation and backfilling.					
L121	In manholes and other chambers: excavation of rock.	m^3	27			
L122	In manholes and other chambers: excavation of mass concrete.	m^3	4			
L141	In thrust boring: excavation of rock.	m^3	127			
	Special pipe laying methods.					
L223	Thrust boring pipe nominal bore 300–600 mm; manholes 18 to 19.	m	98			
				PAGE TOTAL		

Number	Item description	Unit	Quantity	Rate	Amount	
					£	p
	PIPEWORK – SUPPORTS AND PROTECTION ANCILLARIES TO LAYING AND EXCAVATION.					
	Beds depth 150 mm.					
L331	Imported granular material type C2 pipe nominal bore not exceeding 200 mm.	m	27			
L332	Imported granular material type C2 pipe nominal bore 200–300 mm.	m	2483			
L352	Reinforced concrete mix ST3 reinforced as drawing 137/19 detail B pipe nominal bore 200–300 mm.	m	125			
L353	Reinforced concrete mix ST3 reinforced as drawing 137/19 detail A pipe nominal bore 300–600 mm.	m	90			
	Haunches.					
L431	Imported granular material type C2 pipe nominal bore not exceeding 200 mm; bed depth 200 mm.	m	50			
L432	Imported granular material type C2 pipe nominal bore 200–300 mm; bed depth 200 mm.	m	250			
	Beds and surrounds.					
L531.1	Imported granular material type C2 pipe nominal bore not exceeding 200 mm; bed depth 200 mm.	m	402			
	Surrounds.					
531.2	Imported granular material type C2 to two pipes, maximum distance between inside face of outer walls of pipes 600 mm; bed depth 200 mm.	m	41			
L601	Wrapping and lagging pipe nominal bore not exceeding 200 mm with Wrappo.	m	207			
				PAGE TOTAL		

Number	Item description	Unit	Quantity	Rate	Amount	
					£	p
	PIPEWORK – SUPPORTS AND PROTECTION ANCILLARIES TO LAYING AND EXCAVATION.					
	Stools and thrust blocks.					
L721	Volume 0.1–0.2 m^3 concrete grade G21 pipe nominal bore not exceeding 200 mm.	nr	4			
L732	Volume 0.2–0.5 m^3 concrete grade G21 pipe nominal bore 200–300 mm with strap as drawing 3/27 detail 5.	nr	2			
L822	Isolated pipe support; height 1–1.5 m pipe nominal bore 200–300 mm; 1200 mm long hanger, 6 mm galvanised mild steel as drawing 3/27 detail 3.	nr	3			
				PAGE TOTAL		

Number	Item description	Unit	Quantity	Rate	Amount	
					£	p
	STRUCTURAL METALWORK.					
	Conveyor gantry example C, steel grade 43A.					
	Fabrication of members for frames straight on plan.					
M311	Columns.	t	1.4			
M321	Beams.	t	0.9			
M351	Roof trusses comprising single 70 × 70 × 8 angle rafters and 50 × 50 × 6 internal and bottom ties.	t	0.9			
M353	Built-up side girders cambered comprising two single 150 × 90 × 12 angles top boom, two single 150 × 75 × 12 angles bottom boom with verticals 70 × 70 × 8 angles, diagonals 70 × 70 × 8, 80 × 80 × 8 and 90 × 90 × 10 angles.	t	2.6			
M361	Bracings, purlins and cladding rails.	t	1.7			
M370	Grillages.	t	0.4			
M380	Anchorages and holding bolt assemblies comprising 4 nr 450 × 24 bolts with plates 150 × 150 × 10.	nr	4			
	Erection of conveyor gantry.					
M620	Frame members.	t	7.9			
M632	Site bolts black diameter 16–20 mm.	nr	150			
M662	HSFG load indicating bolts diameter 16–20 mm with washers.	nr	84			
	Conveyor gantry example C, steel grade 43A.					
	Off Site surface treatment.					
M810	Blast cleaning to BS 4232 second quality.	m^2	241			
M870	Painting one coat zinc epoxy primer.	m^2	241			
				PAGE TOTAL		

M

Number	Item description	Unit	Quantity	Rate	Amount	
					£	p
	MISCELLANEOUS METALWORK.					
N110.1	Stairways and landings; staircase S3 drawing 136/27.	t	3.7			
N110.2	Stairways and landings; staircase S4 drawing 136/28.	t	4.5			
N130.1	Galvanised mild steel ladders; stringers 65 × 13 mm, 400 mm apart, rungs 20 mm diameter at 300 mm centres; stringers extended and returned 1000 mm to form handrail.	m	14			
N130.2	Galvanised mild steel ladders; stringers 65 × 13 mm, 400 mm apart, rungs 20 mm diameter at 300 mm centres; stringers extended and returned 1000 mm to form handrail; with safety cage of 3 nr 65 × 13 mm verticals and 65 × 13 mm hoops 750 mm diameter at 700 mm centres cage commencing 2500 mm above commencing level.	m	24			
N161.1	Miscellaneous framing; galvanised mild steel angle section 64 × 64 × 9 mm.	m	247			
N161.2	Miscellaneous framing; galvanised mild steel angle section 38 × 38 × 6 mm.	m	53			
N170	Plate flooring; galvanised mild steel chequer plating 10 mm thick as detailed on drawing 136/42.	m^2	86			
N180	Open grid flooring; galvanised mild steel to BS 4592 as detailed on drawing 137/49.	m^2	47			
N230	Duct covers; galvanised mild steel 10 mm thick, width 300 mm as detailed on drawing 137/50.	m^2	27			
N286	Covered tanks volume 100–300 m^3; galvanised mild steel to BS 417 part 1 reference T40.	nr	3			
N999.1	Galvanised mild steel adjustable V-notch weir plate to precise levels as detailed on drawing 136/7 detail A.	m	137			
N999.2	Cast iron light duty inspection cover to BS 497 reference C6-24/24.	nr	2			
N999.3	Supply cast iron step irons to BS 1247 type a2.	nr	12			
N999.4	Supply 2200 × 2200 mm galvanised mild steel forebay screen as detailed on drawing 136/98.	t	1.4			
				PAGE TOTAL		

N

Number	Item description	Unit	Quantity	Rate	Amount £	Amount p
	TIMBER.					
	Hardwood components cross-sectional are 0.01–0.02 m^2; wrought finish.					
O122	150 × 75 mm Greenheart length 1.5–3 m.	m	184			
O123	150 × 100 mm Greenheart length 3–5 m; pier decking runners.	m	204			
	Hardwood components cross-sectional area 0.04–0.1 m^2; wrought finish.					
O143	300 × 300 mm Greenheart length 3–5 m; pier braces between piles.	m	163			
	Hardwood components cross-sectional area 0.1–0.2 m^2; wrought finish.					
O152	350 × 350 mm Greenheart length 1.5–3 m.	m	287			
O153	400 × 400 mm Greenheart length 3–5 m; pier decking bearers.	m	364			
	Softwood components cross-sectional area not exceeding 0.01 m^2; wrought finish.					
O211	75 × 75 mm Douglas fir tanalised length not exceeding 1.5 m.	m	196			
O212	100 × 100 mm Douglas fir tanalised length 1.5–3 m.	m	47			
O213	100 × 50 mm Douglas fir tanalised with rounded edges length 3–5 m; hand rail.	m	280			
	Softwood components cross-sectional area 0.04–0.1 m^2; wrought finish.					
O242	150 × 300 mm Douglas fir length 1.5–3 m.	m	48			
O243	225 × 300 mm Douglas fir length 3–5 m; pier rubbing piece.	m	440			
				PAGE TOTAL		

Number	Item description	Unit	Quantity	Rate	Amount £	Amount p
	TIMBER.					
	Hardwood decking thickness 50–75 mm; wrought finish.					
O330.1	150 × 70 mm Greenheart.	m^2	786			
O330.2	250 × 70 mm Greenheart.	m^2	260			
	Fittings and fastenings.					
	Galvanised mild steel.					
O510	Straps girth: 457 mm width 50 mm thickness 5 mm as drawing D3/27.	nr	10			
O520	Spikes: length 75 mm.	nr	50			
O540.1	Bolts: length 75 mm, diameter 5 mm.	nr	50			
O540.2	Bolts: stainless steel length 100 mm diameter 6 mm.	nr	40			
O550	Plates: stainless steel 100 × 100 mm thickness 6 mm.	nr	10			
				PAGE TOTAL		

Number	Item description	Unit	Quantity	Rate	Amount £	Amount p
	Piling to Pumping Station.					
	Commencing Surface 33.00 a.o.d.					
	Bored cast in place piles concrete grade C25 as Specification Clause 713.3.					
	Diameter 900 mm.					
P151	Number of piles.	nr	97			
P152	Concreted length.	m	1195			
P153	Depth bored maximum depth 18 m.	m	1305			
	Diameter 1200 mm raked at inclination ratio 1 : 6.					
P161	Number of piles.	nr	42			
P162	Concreted length.	m	932			
P163	Depth bored maximum depth 35 m.	m	992			
				PAGE TOTAL		

Number	Item description	Unit	Quantity	Rate	Amount £	Amount p
	Piling to Bridge.					
	Commenced Surface to be Original Surface.					
	Preformed piles concrete grade C25 as Specification Clause 713.4 reinforcement as detail 4 drawing 137/65 circular diameter 300 mm.					
P331.1	Number of piles length 8.5 m; mild steel driving heads and shoes.	nr	10			
P331.2	Number of piles length 12.5 m; mild steel driving heads and shoes.	nr	17			
P332	Depth driven.	m	284			
	Preformed piles concrete grade C25 as Specification Clause 713.4 reinforcement as detail 5 on drawing 137/65 nominal circular diameter 450 mm.					
P351.1	Number of piles length 12.5 m.	nr	128			
P351.2	Number of piles length 17.2 m.	nr	40			
P352	Depth driven.	m	2120			
				PAGE TOTAL		

P

Number	Item description	Unit	Quantity	Rate	Amount £	Amount p
	Piling to Pier.					
	Timber piles cross-sectional area: 0.15–0.25 m^2: 400 × 400 mm Greenheart.					
	Commencing Surface to be Original Surface of bed of River Corve.					
P651	Number of piles length 7.5 m; galvanised mild steel driving head and shoe as detail 5 drawing 140/7.	nr	104			
P652	Depth driven.	m	630			
				PAGE TOTAL		

Number	Item description	Unit	Quantity	Rate	Amount	
					£	p
	PILING TO STORAGE TANK.					
	Interlocking grade 43A steel piles Lincolnshire Steel type 2N section modulus 1150 cm^2/m.					
	Commencing Surface 300 mm below Original Surface.					
P831.1	Length of corner piles.	m	44			
P831.2	Length of taper piles.	m	27			
P832	Driven area.	m^2	6847			
P833	Area of piles of length: not exceeding 14 m; treated with two coats bitumen paint.	m^2	3764			
P834	Area of piles of length: 14–24 m; treated with two coats bitumen paint.	m^2	4367			
				PAGE TOTAL		

Number	Item description	Unit	Quantity	Rate	Amount	
					£	p
	PILING ANCILLARIES.					
	Cast in place concrete piles.					
Q125	Backfilling empty bore with selected, excavated material diameter 900 mm.	m	20			
Q135	Permanent casings each length not exceeding 13 m, diameter 900 mm diameter pile.	m	40			
Q155	Enlarged bases; 2500 mm diameter to 900 mm diameter pile.	nr	87			
Q156	Enlarged bases; 3000 mm diameter to 1200 mm diameter pile.	nr	42			
Q185	Preparing heads; 900 mm diameter.	nr	87			
Q186	Preparing heads; 1200 mm diameter.	nr	42			
	Mild steel reinforcement to BS 4449.					
Q211	Straight bars, nominal size not exceeding 25 mm.	t	46.3			
Q212	Straight bars, nominal size exceeding 25 mm.	t	27.3			
Q700	Obstructions.	h	150			
				PAGE TOTAL		

Q

Number	Item description	Unit	Quantity	Rate	Amount £	Amount p
	PILING ANCILLARIES.					
	Timber piles cross-sectional areas 0.15–0.25 m^2.					
Q415	Pre-boring.	m	200			
Q445	Number of pile extensions.	nr	50			
Q455	Length of pile extensions each length not exceeding 3 m.	m	75			
Q475	Cutting off surplus lengths.	m	25			
Q700	Obstructions.	h	10			
Q811	Pile tests, maintained loading with various reactions test load: 75 t; to working pile.	nr	5			
				PAGE TOTAL		

Number	Item description	Unit	Quantity	Rate	Amount	
					£	p
	PILING ANCILLARIES.					
	Interlocking steel piles section modulus 1150 cm^3/m.					
Q643	Number of pile extensions.	nr	150			
Q653	Length of pile extensions each length: not exceeding 3 m.	m	300			
Q673	Cutting off horizontal surplus lengths, average depth 1.5 m.	m	220			
Q700	Obstructions.	h	20			
				PAGE TOTAL		

Number	Item description	Unit	Quantity	Rate	Amount	
					£	p
	ROADS AND PAVINGS.					
	Sub-bases, flexible road bases and surfacing.					
R118	Granular material DTp Specified type 1 depth 475 mm.	m^2	3760			
R216	Wet-mix macadam DTp Specified Clause 808 depth 210 m.	m^2	3760			
R232	Dense bitumen macadam DTp Specified Clause 903 depth 60 mm.	m^2	3760			
R322	Rolled asphalt DTp Specified Clause 907 depth 40 mm.	m^2	3760			
R341	Surface dressing depth 25 mm; nominal 20 mm coated chippings.	m^2	3760			
	Precast concrete kerbs, to BS 7263: Part 1 figure 1(c) bedded and backed with concrete ST1 cross-section 300 × 350 mm.					
R611	Straight or curved to radius exceeding 12 m.	m	917			
R612	Curved to radius not exceeding 12 m.	m	48			
R613	Quadrants.	nr	27			
R641	Precast concrete channels to BS 7263: Part 1 figure 1(h) straight or curved to radius exceeding 12 m; 400 × 150 mm bed.	m	917			
R642	Precast concrete channels to BS 7263: Part 1: figure 1(h) curved to radius not exceeding 12 m; 400 × 150 mm bed.	m	48			
R643	Precast concrete channels to BS 7263: Part 1: figure 1(h) quadrant; 400 × 150 mm bed.	nr	27			
	Light duty pavements.					
R714	Granular base depth 150 mm; DTp Specified type 1.	m^2	1426			
R727	Hardcore base depth 300 mm.	m^2	387			
R774	In situ concrete grade C15 depth 150 mm.	m^2	387			
R783	Precast concrete flags to BS 7263: Part 1: type D; thickness 63 mm.	m^2	1426			
R900	Joint new concrete road to existing concrete road.	m	27			
				PAGE TOTAL		

R

Number	Item description	Unit	Quantity	Rate	Amount	
					£	p
	ROADS AND PAVINGS.					
	Sub-bases, flexible road bases and surfacing.					
R124	Granular material DTp Specified type 2 depth 150 mm.	m^2	1039			
R170	Geotextiles; Georam, grade G2.	m^2	1039			
R180	Additional depth of hardcore.	m^3	380			
	Concrete pavements.					
R414.1	Carriageway slabs of DTp Specified paving quality concrete depth 150 mm.	m^2	1039			
R414.2	Carriageway slabs of DTp Specified paving quality concrete depth 150 mm; inclined at an angle exceeding 10°.	m^2	1764			
R443	Steel fabric reinforcement to BS 4483 nominal mass 3–4 kg/m^2; type A252.	m^2	1039			
R480	Waterproof membrane below concrete pavements; 500 grade impermeable plastic sheeting.	m^2	1039			
	Joints in concrete pavements.					
R524	Expansion joints depth 100–150 mm; as detail C drawing 137/51 at 5 m centres.	m	321			
R534	Contraction joints depth 100–150 mm; as detail D drawing 137/51 at 2.5 m centres.	m	47			
	Kerbs, channels and edgings.					
R651	Precast concrete edgings to BS 7263: Part 1: figure 1(m) straight or curved to radius exceeding 12 m; 200 × 200 mm concrete ST1 bed and haunch.	m	127			
R652	Precast concrete edgings to BS 7263: Part 1: figure 1(m) curved to radius not exceeding 12 m; 200 × 200 mm concrete ST1 bed and haunch.	m	480			
				PAGE TOTAL		

R

Number	Item description	Unit	Quantity	Rate	Amount £	Amount p
	ROADS AND PAVINGS.					
	Light duty pavements.					
R713	Granular base DTp Specified type 2 depth 100 mm.	m^2	840			
R714	Granular base DTp Specified type 2 depth 150 mm; inclined at an angle exceeding 10° to the horizontal.	m^2	37			
R782.1	Precast concrete flags to BS 7263: Part 1 type D depth 50 mm.	m^2	330			
R782.2	Precast concrete flags to BS 7263: Part 1 type D depth 50 mm; inclined at an angle exceeding 10° to the horizontal.	m^2	390			
				PAGE TOTAL		

R

Number	Item description	Unit	Quantity	Rate	Amount £	Amount p
	RAIL TRACK.					
S110	Track foundation, bottom ballast, crushed granite.	m^3	636			
S120	Track foundation, waterproof membrane, visqueen 1000 gauge.	m^2	1250			
	Taking up track.					
	Welded track on concrete sleepers, fully dismantled and placed in Employer's store at Craven Arms p.w.d.					
S211	Bull head rail, plain track.	m	2000			
S214	Bull head rail, turnouts.	nr	4			
S250	Conductor rails.	m	4000			
S281	Sundries, buffer stops, approximate weight 2.5 tonnes of steel rail and timber sleeper construction.	nr	4			
S283	Sundries, wheelstops.	nr	2			
	Lifting, packing and slewing.					
S310	Bull head rail track, length 20 m, maximum distance of slew 200 mm, maximum lift 100 m.	nr	1			
	Supplying.					
S425	Flat bottom rail, reference 113A, mass 56 kg/m.	t	38			
S471	Sleepers, softwood timber 2600 × 250 × 130 mm.	nr	504			
S472	Sleepers, concrete 2600 × 250 × 150 mm type C2.	nr	27			
S481	Fittings, chairs, type 4.	nr	1008			
S482	Fittings, base plates, type 3.	nr	1008			
				PAGE TOTAL		

Number	Item description	Unit	Quantity	Rate	Amount £	Amount p
	RAIL TRACK.					
	Supplying.					
S483	Pandrol rail fastenings, type 7.	nr	94			
S484	Plain fish plates, type 9.	nr	127			
S510	Turnouts, type T4, drawing 27.	nr	4			
S520	Diamond crossings, type DC1 drawing 27.	nr	2			
S581	Sundries, buffer stops, type B2 approximate weight 2.5 tonnes.	nr	4			
S585	Sundries, switch heaters, type SH2.	nr	6			
	Laying flat bottom rails.					
S621.1	Plain track; rail reference 113A, mass 56 kg/m, fish plated joints on timber sleepers.	m	3200			
S621.2	Plain track; rail reference 113A, mass 56 kg/m, welded joints, on concrete sleepers.	m	2000			
S623	Form curve in plain track, radius exceeding 300 m, welded joints on concrete sleepers.	m	500			
S624	Turnouts type T4, drawing 27, length 26.2 m, fish plated joints on timber sleepers.	nr	4			
S625	Diamond crossings, type DC1 drawing 27, length 27.4 m, fish plated joints on timber sleepers.	nr	2			
S627	Welded joints by Quick Thermit process.	nr	10			
S681	Sundries buffer stops approximate weight 2.5 tonnes.	nr	4			
S685	Sundries switch heaters.	nr	6			
				PAGE TOTAL		

S

Number	Item description	Unit	Quantity	Rate	Amount	
					£	p
	FOREBAY TUNNEL AND OVERFLOW SHAFT.					
	Tunnel excavation diameter 2.5 m.					
T112.1	In rock; straight.	m^3	2180			
T112.2	In rock; curved, material to be used as filling.	m^3	520			
T132	Shaft excavation diameter 2.8 m in rock; straight.	m^3	280			
T170	Excavated surfaces in rock; voids filled with cement grout as Specification Clause 137/T17.	m^2	4352			
T232.1	In situ reinforced cast concrete primary straight tunnel lining internal diameter 2.0 m; concrete as Specification Clause 137/F3.	m^2	820			
T232.2	In situ reinforced concrete primary curved tunnel lining internal diameter 2.0 m; concrete as Specification Clause 137/F3.	m^3	175			
T252.1	In situ straight lining formwork finish grade T2 internal diameter 2.0 m.	m^2	3566			
T252.2	In situ curved lining formwork finish grade T2 internal diameter 2.0 m.	m^2	760			
				PAGE TOTAL		

Number	Item description	Unit	Quantity	Rate	Amount	
					£	p
	DIVERSION TUNNEL.					
	Work to be executed under compressed air at gauge pressure not exceeding one bar.					
T128.1	Excavation straight tunnel diameter 10.3 m in Group 5 material.	m^3	41 000			
T128.2	Excavation straight tunnel diameter 10.3 m in Group 6 material.	m^3	8800			
T180.1	Excavated surface Group 5 materials; voids filled with cement grout as Specification Clause 137/T17.	m^2	15 800			
T180.2	Excavated surface in Group 6 material; voids filled with cement grout as Specification Clause 137/T17.	m^2	3400			
T538	Cast iron bolted segmental tunnel lining rings internal diameter 9.5 m; nominal width 450 mm comprising 16 segments maximum piece weight 110 kg 136 bolts and grummets and 272 washers.	nr	1330			
T571	Parallel circumferential packing for preformed segmental tunnel linings, internal diameter 9.5 m; bitumen impregnated fibreboard thickness 8 mm.	nr	1329			
T574	Leaf fibre caulking for preformed segmental tunnel linings, internal diameter 9.5 m.	m	51 500			
				PAGE TOTAL		

T

Number	Item description	Unit	Quantity	Rate	Amount £	Amount p
	FOREBAY TUNNEL AND OVERFLOW SHAFT.					
	In situ lining to non-circular shafts; rectangular 3.2 × 2.4 m.					
T332	Cast concrete size 2.75 × 1.95 m; concrete as Specification Clause 137/F3.	m^3	130			
T352	Formwork finish Grade T1 size 2.75 × 1.95 m.	m^2	615			
	Support and stabilisation.					
T811	Rock bolts impact expanding mechanical 55 mm diameter with 30 mm square shank maximum length 5 m.	m	108			
	Pressure grouting.					
T831	Sets of drilling and grouting plant.	nr	1			
T832	Face packers.	nr	180			
T834	Drilling and flushing diameter 50 mm length 5–10 m.	m	1080			
T835	Re-drilling and flushing holes length 5–10 m.	m	540			
T836	Injection of cement grout as Specification Clause 137/T801.	t	85			
T840	Forward probing length 10–15 m.	m	480			
				PAGE TOTAL		

T

Number	Item description	Unit	Quantity	Rate	Amount £	p
	BRICKWORK, BLOCKWORK AND MASONRY.					
	Common brickwork to BS 3921: stretcher bond, flush pointed mortar type M2.					
U111	102.5 mm nominal thickness vertical straight walls; cavity construction.	m^2	287			
U160	Columns and piers, cross-sectional dimensions 600 × 600 mm.	m	43			
	Ancillaries.					
U182.1	Damp proof courses; width 100 mm to BS 743 type D.	m	150			
U182.2	Damp proof courses; width 225 mm to BS 743 type D.	m	17			
U185	Concrete infill grade 1; thickness 50 mm.	m^2	31			
U186	Fixing and ties; galvanised mild steel in accordance with Specification Clause 5/27.	m^2	304			
U187	Built-in pipes and ducts cross-sectional area not exceeding 0.05 m^2.	nr	69			
U188	Built-in pipes and ducts cross-sectional area 0.10 m^2.	nr	6			
	Facing brickwork.					
	Plowden Red facings as Specification Clause 5/29.					
U211	102.5 mm nominal thickness vertical straight wall; stretcher bond flush pointed, mortar type M3 cavity construction.	m^2	321			
	Surface features.					
U271.1	Brick on edge coping.	m	143			
U271.2	Special sill as detail D drawing 137/97.	m	37			
				PAGE TOTAL		

Number	Item description	Unit	Quantity	Rate	Amount £	Amount p
	BRICKWORK. BLOCKWORK AND MASONRY.					
	Brickwork ancillaries.					
U282	Damp proof courses; width 100 mm to BS 743 type D.	m	187			
U287	Built-in pipes and ducts cross-sectional area not exceeding 0.05 m^2.	nr	69			
U288	Built-in pipes and ducts cross-sectional area 0.10 m^2.	nr	6			
	Engineering brickwork class B to BS 3921, stretcher bond, flush pointed mortar type M4.					
U333	450 mm nominal thickness battered straight wall.	m^2	87			
U334	450 mm nominal thickness battered curved wall.	m^2	184			
U371	Bullnosed copings.	m	47			
U375	Bullnosed corbels.	m	83			
U383	Movement joint; 25 mm bitumen impregnated fibreboard, mean width 450 mm with plastijoint sealer both sides.	m	27			
U384	Bond to existing work.	m^2	14			
	Lightweight blockwork. hollow block to BS 6073, stretcher bond, flush pointed mortar type M1.					
U411.1	Vertical 100 mm nominal thickness straight wall.	m^2	203			
U411.2	Vertical 140 mm nominal thickness straight wall.	m^2	127			
U482.1	Damp proof courses; width 100 mm to BS 743 type D.	m	83			
U482.2	Damp proof courses; width 140 mm to BS 743 type D.	m	46			
				PAGE TOTAL		

Number	Item description	Unit	Quantity	Rate	Amount £	Amount p
	BRICKWORK. BLOCKWORK AND MASONRY.					
	Dense concrete blockwork, solid block to BS 6073, stretcher bond, flush pointed mortar type M1.					
U511	Vertical 140 mm nominal thickness straight wall.	m^2	83			
U582	Damp proof courses; width 140 mm to BS 743 type D.	m	27			
U587	Built-in pipes and ducts cross-sectional area not exceeding 0.05 m^2.	nr	41			
	Ashlar masonry Portland stone rubbed finish Specification Clause 5/40 mortar type M2.					
U735	300 mm nominal thickness vertical facing to concrete flush pointed.	m^2	4387			
U771	Surface features; rounded copings 1200 × 600 mm as drawing 137/91.	m	1236			
U786	Ancillaries fixings and ties; Alleyslots at 1500 mm centres.	m^2	4387			
U787	Ancillaries built-in pipes and ducts cross-sectional area 0.025–0.25 m^2.	nr	45			
U799	Bollards diameter 600 mm height 900 mm as drawing 137/93.	nr	24			
				PAGE TOTAL		

U

Number	Item description	Unit	Quantity	Rate	Amount £	Amount p
	PAINTING.					
V113	Zinc rich primer paint on metal surfaces other than metal sections and pipework inclined at an angle exceeding 60° to the horizontal; in one coat.	m^2	47			
V116	Zinc rich primer paint on metal surfaces other than metal sections and pipework width not exceeding 300 mm; in one coat.	m	91			
V118	Zinc rich primer paint on metal surfaces other than metal sections and pipework isolated groups of surfaces; 300 × 200 mm inspection covers and frames in one coat.	nr	14			
	Oil paint in three coats.					
V313	On metal surfaces other than metal sections and pipework; inclined at an angle exceeding 60° to the horizontal.	m^2	47			
V316	On metal surfaces other than metal sections and pipework of width not exceeding 300 mm.	m	91			
V318	On metal surfaces other than metal sections and pipework isolated groups of surfaces; 300 × 200 mm inspection covers and frames.	nr	14			
V326	Timber surfaces of width not exceeding 300 mm.	m	14			
	Emulsion paint in three coats.					
V533	Smooth concrete surfaces inclined at an angle exceeding 60° to the horizontal.	m^2	487			
V534	Smooth concrete soffit and lower surfaces inclined at an angle not exceeding 60° to the horizontal.	m^2	326			
V563	Brickwork and blockwork surfaces inclined at an angle exceeding 60° to the horizontal.	m^2	3867			
V566	Brickwork and blockwork surfaces of width not exceeding 300 mm.	m	187			
V633	Cement paint on smooth concrete surfaces inclined at an angle exceeding 60° to the horizontal; in one coat.	m^2	27			
V636	Cement paint on smooth concrete surfaces of width not exceeding 300 mm; in one coat.	m	14			
				PAGE TOTAL		

Number	Item description	Unit	Quantity	Rate	Amount £	Amount p
	WATERPROOFING.					
W131	Damp proofing, waterproof sheeting to upper surfaces inclined at an angle not exceeding 30° to the horizontal; polythene to BS 3012 in one layer thickness 5 mm.	m^2	320			
W211	Tanking, asphalt to upper surfaces inclined at an angle not exceeding 30° to the horizontal; mastic asphalt to BS 1097 in one coating thickness 25 mm.	m^2	410			
W213	Tanking, asphalt to surfaces inclined at an angle exceeding 60° to the horizontal; mastic asphalt to BS 1097 in one coating thickness 25 mm.	m^2	534			
W311	Roofing asphalt upper surfaces inclined at an angle not exceeding 30° to the horizontal; mastic asphalt to BS 988 in two coatings total thickness 25 mm including heavy gauge polythene isolating membrane non-staining roofing felt to BS 747 type 1C expanded polystyrene thickness 25 mm and concrete grade 10 screened average thickness 75 mm.	m	726			
W441	Protective layer sand and cement screed type A thickness 25 mm upper surfaces inclined at an angle not exceeding 30° to the horizontal.	m^2	410			
				PAGE TOTAL		

Number	Item description	Unit	Quantity	Rate	Amount £	Amount p
	WATERPROOFING.					
	Roofing.					
W371	Profiled aluminium sheet to BS 4868 type A 1 mm thick in one layer upper surfaced inclined at an angle not exceeding 30° to the horizontal.	m^2	463			
W373	Profiled aluminium sheet to BS 4868 type A 1 mm thick in one layer surfaces inclined at an angle exceeding 60° to the horizontal.	m^2	125			
W371	Waterproof sheeting upper surfaces inclined at an angle not exceeding 30° to the horizontal; corrugated plastic translucent sheet to BS 4154 1.5 mm thick.	m^2	57			
				PAGE TOTAL		

Number	Item description	Unit	Quantity	Rate	Amount	
					£	p
	MISCELLANEOUS WORK.					
	Fences.					
X113	Timber post and rail fence height 1.3 m; to BS 1722 Part 7.	m	187			
X133	Concrete post and wire plastic coated chain link fence to BS 1722 Part 1 height 1.4 m concrete grade 10 foundations 450 × 450 × 450 mm deep.	m	470			
X136	Concrete post and wire anti-intruder chain link fence to BS 1722 Part 10 height 2.9 m; with cranked posts concrete grade 10 foundations 600 × 600 × 750 mm deep.	m	305			
X163.1	Timber close boarded fence height 1.4 m; with concrete posts to BS 1755 Part 5 concrete grade 10 foundations on surface inclined at an angle exceeding 10°.	m	13			
X191	Chestnut pale fence height 900 mm; with timber posts to BS 1722 Part 4.	m	403			
X193	Mild steel unclimbable fence height 1.372 m; to BS 1722 Part 9 concrete grade 10 foundations 450 × 450 × 450 mm deep.	m	108			
X198	Mild steel crash barrier height 600 mm concrete ST1 foundations 300 × 300 × 450 mm deep.	m	1871			
				PAGE TOTAL		

X

Number	Item description	Unit	Quantity	Rate	Amount £	p
	MISCELLANEOUS WORK.					
	Gates and stiles.					
X215	Timber field gate, 2 leaves, overall width 3.353 m height 1.143 m; to BS 3470.	nr	14			
X235	Metal field gate, width 3.353 m height 1.143 m; to BS 3470.	nr	2			
X295	Entrance gate, width 3.5 m height 1.4 m; galvanised mild steel angle sections filled in with plastic coated chain link fencing as detail D drawing 137/75.	nr	1			
	Drainage to structures above ground unplasticised PVC to BS 4576.					
X331.1	Gutters; 100 mm diameter.	m	47			
X331.2	Gutters; 150 mm diameter.	m	36			
X332.1	Gutter bends 90°; 100 mm diameter.	nr	10			
X332.2	Gutter outlets; 150 mm diameter.	nr	8			
X333.1	Downpipes; 100 mm diameter.	m	23			
X333.2	Downpipes; 150 mm diameter.	m	16			
X334.1	Swan necks; 100 mm diameter.	nr	4			
X334.2	Shoes; 150 mm diameter.	nr	3			
				PAGE TOTAL		

Number	Item description	Unit	Quantity	Rate	Amount	
					£	p
	MISCELLANEOUS WORK.					
	Rock filled gabions.					
X410.1	Box gabion 1 × 1 × 2 m: 4 mm × 100 × 100 mm galvanised wire mesh, excavated rock grade GB1.	nr	27			
X410.2	Box gabion 1 × 1 × 2 m: 4 mm × 100 × 100 mm galvanised wire mesh excavated rock grade GB1.	nr	45			
X420	Mattress gabion, thickness 150 mm: 4 mm × 100 × 100 mm galvanised wire mesh, excavated rock grade GB2.	m^2	50			
				PAGE TOTAL		

Number	Item description	Unit	Quantity	Rate	Amount £	Amount p
	NEWTON STREET MANHOLES 25–30.					
	EXISTING BRICK SEWER NOMINAL SIZE 600 mm OVOID.					
	Preparation.					
Y110	Cleaning.	m	57			
	Removing intrusions.					
Y121	Lateral; bore not exceeding 150 mm; clayware.	nr	12			
Y123	Dead water main; bore not exceeding 150 mm; cast iron.	nr	1			
Y130	CCTV Survey.	m	57			
	Plugging laterals with grout as Specification Clause 7.21.					
Y131	Bore not exceeding 300 mm.	nr	21			
Y132	U-shaped; internal cross-sectional dimensions 350 mm × 200 mm.	nr	2			
	Filling laterals with grout as Specification clause 7.21.					
Y142	Internal cross-section dimensions 450 mm × 350 mm U-shaped.	m^3	1			
	Local internal repairs.					
Y151	Area not exceeding 0.1 m^2.	nr	6			
Y152	Area 0.1–0.25 m^2.	nr	4			
Y153	Area 0.6 m^2.	nr	1			
				PAGE TOTAL		

Number	Item description	Unit	Quantity	Rate	Amount	
					£	p
	EXISTING BRICK SEWER NOMINAL SIZE 600 mm OVOID.					
	Stabilisation of existing sewers.					
Y210	Pointing brickwork with cement mortar.	m^2	14			
Y290	Pointing pipe joints under pressure with epoxy mortar.	m^2	7			
	External grouting.					
Y231	Number of holes.	nr	4			
Y232	Injection of cement grout as Specification Clause 7.24.	m^3	6			
	Renovation of existing sewers.					
	Sliplinings.					
Y311	Butt fusion welded HDPE type SDR 4, thickness 4 mm, 500 mm minimum internal diameter.	m	57			
				PAGE TOTAL		

Number	Item description	Unit	Quantity	Rate	Amount £	Amount p
	EXISTING BRICK SEWER NOMINAL SIZE 1200 × 900 mm, EGG SHAPED.					
	Renovation of existing sewers.					
	Segmental linings.					
Y333	Glass reinforced plastic as Specification Clause 7.50, minimum internal cross-section dimensions 1050 × 750 mm.	m	45			
Y334.1	Glass reinforced concrete, 15 mm thick, internal cross-section dimensions 1050 × 750 mm, egg shaped.	m	38			
Y334.2	Glass reinforced concrete, 15 mm thick, internal cross-section dimensions 1320 × 850 mm, curved to offset of 70 mm per metre.	m	2			
	Annulus grouting.					
Y360	Cement grout as Specification Clause 7.25.	m^3	10			
	Laterals to renovated sewers.					
	Jointing.					
Y411.1	Bore: not exceeding 150 mm, to HDPE sliplining, type SDR 17.	nr	10			
Y411.2	Bore: 150–300 mm, to HDPE sliplining, type SDR 17, regarded.	nr	3			
Y413	900 × 600 mm, egg shaped, to GRP segmental lining, 15 mm thick.	nr	1			
				PAGE TOTAL		

Number	Item description	Unit	Quantity	Rate	Amount £	p
	EXISTING BRICK SEWER NOMINAL SIZE 1200 × 900 mm, EGG SHAPED.					
	Renovation of existing sewers.					
	Segmental linings.					
Y333	Glass reinforced plastic as Specification Clause 7.50, minimum internal cross-section dimensions 1050 × 750 mm.	m	45			
Y334.1	Glass reinforced concrete, 15 mm thick, internal cross-section dimensions 1050 × 750 mm, egg shaped.	m	38			
Y334.2	Glass reinforced concrete, 15 mm thick, internal cross-section dimensions 1320 × 850 mm, curved to offset of 70 mm per metre.	m	2			
	Annulus grouting.					
Y360	Cement grout as Specification Clause 7.25.	m^3	10			
	Laterals to renovated sewers.					
	Jointing.					
Y411.1	Bore: not exceeding 150 mm, to HDPE sliplining, type SDR 17.	nr	10			
Y411.2	Bore: 150–300 mm, to HDPE sliplining, type SDR 17, regarded.	nr	3			
Y413	900 × 600 mm, egg shaped, to GRP segmental lining, 15 mm thick.	nr	1			
				PAGE TOTAL		

Number	Item description	Unit	Quantity	Rate	Amount £	Amount p
	EXISTING BRICK SEWER NOMINAL SIZE 1200 × 900 mm, EGG SHAPED.					
	Flap valves.					
Y421.1	Remove existing; nominal diameter 225 mm.	nr	2			
Y421.2	Remove existing; nominal diameter 300 mm.	nr	3			
Y422	Replace existing; nominal diameter 225 mm.	nr	2			
Y423	New; Holdwater catalogue reference 918; nominal diameter 300 mm.	nr	3			
	New manholes.					
Y625	Brick with backdrop type 1A, heavy duty cover and frame; depth 3–3.5 m.	nr	2			
Y657	Precast concrete type 2B, heavy duty cover and frame; depth 7.5 m.	nr	1			
	New manholes replacing existing manholes.					
Y653	Precast concrete type 2C, heavy duty cover and frame; depth 2–2.5 m.	nr	2			
				PAGE TOTAL		

Number	Item description	Unit	Quantity	Rate	Amount £	Amount p
	VALVES SV 17-32.					
	EXISTING CAST IRON MAINS NOMINAL BORE NOT EXCEEDING 200–300 mm.					
Y512	Cleaning.	m	60			
Y522	Removing intrusions.	nr	15			
Y542	CCTV Survey.	m	60			
Y562	Epoxy lining as Specification Clause 263.07.	m	60			
				PAGE TOTAL		

Number	Item description	Unit	Quantity	Rate	Amount £	Amount p
	EXISTING BRICK SEWER NOMINAL SIZE 1200 × 900 mm, EGG SHAPED.					
	Existing manholes.					
	Abandonment, as drawing 22/C.					
Y713	Depth 2–2.5 m including removing cover slab, breaking back shaft and backfilling with pulverised fuel ash.	nr	1			
Y717	Depth 5.5 m including removing cover slab, breaking shaft and backfilling with pulverised fuel ash.	nr	1			
	Alterations.					
Y720	Work to benching and inverts, as drawing 27/G including breading out, re-haunching and dealing with flows.	nr	1			
	Interruptions.					
Y810	Preparation of existing sewers.	h	10			
Y820	Stabilisation of existing sewers.	h	10			
	Renovation of existing sewers.					
Y833	Segmental linings.	h	20			
				PAGE TOTAL		

Number	Item description	Unit	Quantity	Rate	Amount	
					£	p
	GATE HOUSE.					
	EARTHWORKS.					
E***	(measured in accordance with class E)					
	CONCRETE.					
F–H***	(measured in accordance with classes F–H)					
	METAL CLADDING.					
N210	(measured in accordance with class N)					
	BRICKWORK, BLOCKWORK AND MASONRY.					
U***	(measured in accordance with class U)					
	CARPENTRY AND JOINERY.					
	Sawn softwood structural and carcassing timber as Specification Clause 24.07.					
Z112.1	In walls and partitions; 50 × 38 mm.	m	22			
Z112.2	In walls and partitions; 75 × 50 mm.	m	36			
Z114.1	In pitched roofs; 75 × 50 mm.	m	48			
Z114.2	In pitched roofs; 100 × 50 mm.	m	108			
Z114.3	In pitched roofs; 150 × 25 mm.	m	18			
Z132	Chipboard tongued and grooved floor boarding 19 mm thick.	m^2	20			
	Miscellaneous joinery.					
Z151	Wrought softwood skirtings, finished size 150 × 19 mm, in two cross-section shapes.	m	18			
Z152	Wrought softwood architraves, finished size 46 × 19 mm, in two cross-section shapes.	m	15			
				PAGE TOTAL		

Z

Number	Item description	Unit	Quantity	Rate	Amount	
					£	p
	CARPENTRY AND JOINERY.					
	Miscellaneous joinery.					
Z153.1	Wrought softwood beading, finished size 25 × 12 mm.	m	8			
Z153.2	Wrought softwood capping, finished size 144 × 19 mm.	m	2			
Z154	Wrought softwood shelves complete with all framing and support, 300 × 12 mm as Drawing No C1/GH/003.	m	6			
	Kitchen units as "Aurora" range manufactured by Kitchens Delight, Unit 7, Stanley Industrial Estate, Cheadle Hulme SK8 7JJ.					
Z161.1	1000 × 600 mm base unit.	nr	1			
Z161.2	1000 × 600 mm sink unit.	nr	1			
Z162	500 × 600 mm wall unit.	nr	2			
Z163	2000 × 600 mm work top.	nr	1			
Z169	1500 × 500 mm single bowl and drainer sink unit.	nr	1			
Z164	Fibreboard notice board 2 m × 1 m × 12 mm thick.	nr	1			
	INSULATION.					
Z221	Fibreglass guilt, 150 mm thick, laid between ceiling joists.	m^2	20			
	WINDOWS, DOORS AND GLAZING.					
	Wrought softwood windows as Drawing CA/GH/06.					
Z311.1	Overall size of opening 300 × 900 mm.	nr	1			
Z311.2	Overall size of opening 1500 × 1000 mm.	nr	1			
Z311.3	Overall size of opening 1500 × 2000 mm.	nr	1			
Z311.4	Overall size of opening 1900 × 2000 mm.	nr	2			
Z313.1	Wrought hardwood door set type 1A as Drawing CA/GH/07, kept clean for staining, size 900 × 1850 mm.	nr	1			
Z313.2	Wrought softwood door set type 1B as Drawing CA/GH/07, size 900 × 1850 mm.	nr	2			
				PAGE TOTAL		

Number	Item description	Unit	Quantity	Rate	Amount £	Amount p
	WINDOWS, DOORS AND GLAZING.					
	Ironmongery as Drawing CA/GH/07.					
Z341.1	Pair of steel butt hinges; 75 × 50 mm.	nr	5			
Z341.2	Pair of steel butt hinges; 100 × 75 mm.	nr	1			
Z342	"Steptoe" door closers, 140 × 22 mm.	nr	3			
Z343	149 × 70 mm horizontal mortice latch.	nr	1			
Z345.1	Handles; ref H1.	nr	1			
Z345.2	Handles; ref H2.	nr	2			
	GLAZING.					
Z351.1	6 mm clear float glass, fixed with putty.	m^2	13			
Z351.2	6 mm patterned glass as Drawing CA/GH/06.	m^2	1			
Z351.3	6 mm Georgian wired glass in vision panels 130 × 850 mm fixed with wood beads.	m^2	1			
Z355	6 mm mirror, 600 × 600 mm with 4 No drilled holes, fixed with dome head mirror screws.	nr	1			
	SURFACE FINISHES. LININGS AND PARTITIONS.					
	In situ finishes, beds and backings.					
Z411	Sand and cement floor screed 38 mm thick, floated finish.	m^2	20			
	Walls.					
Z413.1	3 mm plaster veneer coat.	m^2	90			
Z413.2	Sand and cement backing, 6 mm thick.	m^2	4			
Z413.3	2 coat plaster 13 mm thick.	m^2	12			
Z415	2 coat plaster 13 mm thick to surfaces not exceeding 300 mm wide.	m	3			
				PAGE TOTAL		

Number	Item description	Unit	Quantity	Rate	Amount £	Amount p
	SURFACE FINISHES, LININGS AND PARTITIONS.					
	Quarry tiles as manufactured by Owen Brothers of Wrexham, Clwyd.					
Z421.1	150 × 150 × 16 mm thick to floors.	m^2	6			
	Carpet tiles, Soporo range as supplied by Kingsland Carpets, St Albans, Herts.					
Z421.2	500 × 500 squares to floors.	m^2	12			
	White ceramic tiles as manufactured by Owen Brothers of Wrexham, Clwyd.					
Z423	100 × 100 × 6 mm thick to walls.	m^2	4			
	Vinyl sheet floor covering, "blue heaven" as supplied by Kingsland Carpets, St Albans, Herts.					
Z431	4 mm thick.	m^2	4			
	Plasterboard dry linings to walls and attached columns.					
Z443.1	12 mm thick fixed with plaster dabs.	m^2	45			
Z443.2	12 mm thick, fixed with nails.	m^2	21			
Z445	12 mm thick, foil backed, fixed with plaster dabs, to surfaces not exceeding 300 mm wide.	m	3			
Z452	"Welcome 2000" suspended ceiling to a 600 × 600 mm grid, 150–500 mm deep with 6 No "Centurian" integral light fittings each 300 × 1200 mm.	m^2	20			
Z480	"Hogben" standard single L-shaped WC cubicle partition comprising side panel and door complete with all fittings and ironmongery.	nr	1			
				PAGE TOTAL		

Z

Number	Item description	Unit	Quantity	Rate	Amount	
					£	p
	PIPED BUILDING SERVICES.					
	Cold water installation.					
Z511.1	Copper pipes 12 mm bore.	m	25			
Z511.2	Copper pipes 18 mm bore.	m	15			
	Copper pipe fittings.					
Z512.1	12 mm capillary bends.	nr	12			
Z512.2	12 mm capillary junctions and branches.	nr	8			
Z512.3	18 mm capillary bends.	nr	10			
Z512.4	18 mm capillary taper.	nr	2			
Z512.5	18 mm compression bends.	nr	6			
Z512.6	18 mm compression junctions and branches.	nr	8			
Z512.7	18 mm compression valves.	nr	2			
	Overflow pipework installation.					
Z511.3	12 mm UPVC pipes.	m	3			
	12 mm UPVC solvent welded pipe fittings.					
Z512.8	Bends.	nr	6			
	Preformed pipework insulation 15 mm thick.					
Z513.1	To pipe bore 12 mm.	m	25			
Z513.2	To pipe bore 18 mm.	m	15			
	Boiler plant and ancillaries.					
Z521	5L electric water boiler as Specification Clause 24.15.	nr	1			
Z524	Moulded plastic 50L storage tank.	nr	1			
	Sanitary appliances and fittings as Dyford's "Simplicity" range.					
Z530.1	WC suite with pan, cistern, flush pipe and seat, catalogue ref 164.	nr	1			
Z530.2	Bowl urinal with exposed fixing lugs, cistern, flush pipe, waste fitting and trap, catalogue ref 178.	nr	1			
Z530.3	Wash hand basin with brackets, pillar taps, waste outlet, plug and chain, and trap, catalogue ref 182.	nr	1			
				PAGE TOTAL		

Number	Item description	Unit	Quantity	Rate	Amount £	Amount p
	CABLED BUILDING SERVICES.					
Z711	PVC insulated, 3 core, 12 mm cables laid or drawn in conduits.	m	40			
Z713.1	PVC insulated, 3 core, 12 mm cables fixed to surfaces.	m	12			
Z713.2	Telephone cable.	m	15			
	Cable and conduit final circuits.					
Z772.1	Lighting.	nr	2			
Z772.2	Heating and power.	nr	1			
Z772.3	Motor.	nr	1			
Z772.4	Water heater.	nr	1			
	Equipment.					
Z781	"Vent a quick" extract fan catalogue ref 11.3.	nr	1			
Z711	**Fittings.**					
	Switches.					
Z782.1	Distribution board.	nr	1			
Z782.2	Light switches; one gang.	nr	2			
Z782.3	Light switches; two gang.	nr	1			
Z783	Light fittings.	nr	6			
	Switched power sockets.					
Z784.1	Single.	nr	4			
Z784.2	Double.	nr	2			
Z782.3	Telephone sockets.	nr	2			
	PAINTING.					
V***	(measured in accordance with class V).					
	WATERPROOFING.					
W***	(measured in accordance with class W).					
	DRAINAGE TO STRUCTURES ABOVE GROUND.					
X***	(measured in accordance with class X).					
				PAGE TOTAL		